Lecture Notes
in Control and Information Sciences

195

Editor: M. Thoma

Derong Liu and Anthony N. Michel

Dynamical Systems with Saturation Nonlinearities

Analysis and Design

Springer-Verlag London Ltd.

Authors

Derong Liu
Electrical and Electronics Research Department, General Motors NAO R & D Center,
30500 Mound Road, Box 9055, Warren, Michigan 48090, USA

Anthony N. Michel
Department of Electrical Engineering, University of Notre Dame, Notre Dame,
Indiana 46556, USA

ISBN 978-3-540-19888-8

British Library Cataloguing in Publication Data
Dynamical Systems with Saturation Nonlinearities:Analysis and Design. -
(Lecture Notes in Control & Information Sciences)
 I. Liu, Derong II. Michel, Anthony N. III. Series
 515.352
ISBN 978-3-540-19888-8

Library of Congress Cataloging-in-Publication Data
Liu, Derong, 1963-
 Dynamical systems with saturation nonlinearities : analysis and
design / Derong Liu and Anthony N. Michel.
   p.   cm. -- (Lecture Notes in control and information sciences; 195)
 Includes bibliographical references and index.
 ISBN 978-3-540-19888-8   ISBN 978-3-540-39333-7 (eBook)
 DOI 10.1007/978-3-540-39333-7
   1. Differentiable dynamical systems. 2. Nonlinear theories. 3. Control theory. 4. Digital filters
(Mathematics) 5. Neural networks (Computer science) I. Michel, Anthony N. II. Title. III. Series.
 QA614.8.L58 1994   94-12280
 003'.85--dc20   CIP

© Springer-Verlag London 1994
Originally published by Springer-Verlag London Limited in 1994

Typesetting: Camera ready by authors

69/3830-543210 Printed on acid-free paper

PREFACE

In modern technology, nonlinear system models have assumed an increasingly important role. A class of nonlinearities which is pervasive in the modeling process consists of saturation type nonlinearities. Frequently, such nonlinearities arise because a system is operated outside of its linear range. However, there are also many examples of systems which derive their intended performance from the properties of saturation nonlinearities.

The purpose of the present monograph is to present some new qualitative results for systems endowed with saturation type nonlinearities in three diverse areas: control systems, digital filters, and neural networks. Our presentation is organized into three parts, each of which addresses the above areas in the indicated order. Each part contains an introductory chapter and altogether, there are eleven chapters, including a chapter (the first chapter) which serves as an introduction to the entire monograph. A brief outline of the seven core chapters follows.

In *Chapter 3*, we establish new results for the global asymptotic stability of the null solution of continuous-time and discrete-time finite dimensional systems with state saturation nonlinearities.

In *Chapter 4*, we establish sufficient conditions for the null controllability of a class of discrete-time systems with state saturation and control constraints. Our motivation for this research is that a system of this type may not be null controllable even when its corresponding linear system is stable and controllable. Our results state that if the system is stabilizable by a constrained controller and if the corresponding linear system is controllable, then, the system is null controllable in the sense that any initial state can be steered to the origin in a finite number of steps using the constrained controller. In our results, if a system satisfies the conditions for null controllability, we have determined a linear state feedback controller which stablizes the system. We include specific examples to demonstrate the applicability of the present results.

In *Chapter 6*, we establish new conditions for the non-existence of limit cycles in n^{th} order state-space fixed-point digital filters. We review some of the existing criteria for the absence of limit cycles in fixed-point digital filters and we demonstrate that the present results are easier to apply and are less conservative than corresponding existing results. When applied to second-order fixed-point digital filters, the present results reduce to conditions which are expressed in terms of the components of the coefficient matrix. The generalized overflow nonlinearity employed herein constitutes a generalization of the usual types of overflow arithmetic used in practice. We also develop algorithms which use linear programming techniques and which can be used to determine a required positive definite matrix in the applications of our results.

In *Chapter 7*, by utilizing the Second Method of Lyapunov, we establish sufficient conditions for the global asymptotic stability of the trivial solution of two-dimensional quarter plane state-space digital filters which are endowed with the generalized overflow nonlinearities. Our results in this chapter constitute further developments to the results established in Chapters 3 and 6. We consider the pth power of the l_p vector norm, the l_∞ norm, and the quadratic form as special forms of Lyapunov functions. Our result concerning the quadratic form Lyapunov functions constitutes (to the best of our knowledge) the least conservative criterion for testing the global asymptotic stability of two-dimensional digital filters with overflow nonlinearities and for testing the non-existence of limit cycles in these filters. We generalize our results to the multidimensional case and we demonstrate the applicability of the present results by means of several specific examples.

In *Chapter 9*, we establish results for the analysis and synthesis of a class of feedback neural networks without any restrictions on the interconnecting structure. The class of neural networks considered herein has the basic structure of the analog Hopfield neural network model and utilizes linear saturation functions to represent the neurons. Our analysis results make it possible to identify and locate all the equilibrium points and to determine the qualitative properties of these equilibrium points for a given neural network, in a systematic manner. Our synthesis procedures make it possible to design in a systematic manner neural networks (for associative memories) which store any set of desired (bipolar) patterns as memory vectors.

In *Chapter 10*, we generalize the results of Chapter 9 to develop design procedures for neural networks with sparse interconnecting structure. Our results guarantee that the synthesized neural networks have (arbitrarily) predetermined partial or sparse interconnecting structures and store any set of desired bipolar patterns as memory vectors. We show that a sufficient condition for the existence of solutions for a sparse neural network design is self-feedback for every neuron in the network. We apply our synthesis procedure to the design of cellular neural networks for associative memories. We believe that this work constitutes the first successful synthesis approach for associative memories by means of artificial neural networks with arbitrarily prespecified sparsity constraints for the network's interconnecting structure.

In *Chapter 11*, we conduct an analysis of the robustness properties of a class of neural networks with applications to associative memories. Specifically, for a neural network with nominal parameters which stores a set of desired bipolar memories, we establish sufficient conditions under which the same set of desired bipolar memories is also stored in the network with perturbed parameters. This result enables us to develop a synthesis procedure for associative memories via neural networks whose stored memories are invariant under perturbations and whose upper bounds for allowable parameter perturbations can be chosen beforehand. Our synthesis procedure is capable of generating artificial neural networks with prespecified sparsity constraints (on the interconnecting structure) and with nonsymmetric or symmetric connection matrix.

The research reported in this monograph was supported in part by the National Science Foundation under Grant ECS 91-07728. We are grateful to our colleagues Dr. P. J. Antsaklis, Dr. P. H. Bauer, Dr. M. D. Lemmon, Dr. M. K. Sain, and Dr. K. Wang, Department of Electrical Engineering, for reading different parts of the manuscript and for valuable suggestions.

Derong Liu   Anthony N. Michel
Warren, Michigan   Notre Dame, Indiana
November, 1993   November, 1993

LIST OF FIGURES

CONTENTS

CHAPTER 1

INTRODUCTION TO DYNAMICAL SYSTEMS WITH SATURATION NONLINEARITIES

1.1 Dynamical Systems and Saturation Nonlinearities

The theory of linear systems is mature and well developed. Many problems in linear systems theory have been solved using readily available mathematical tools. On the other hand, the theory of nonlinear systems is not so well developed because more sophisticated mathematical tools are required than in the case of linear systems. The study of nonlinear systems is important because almost every real system contains nonlinearities of one form or another. The primary goal of this monograph is to contribute to the *qualitative theory* of a class of deterministic, finite-dimensional nonlinear dynamical systems.

Studies of nonlinear systems usually address aspects enumerated below.

• *Qualitative Properties of Nonlinear Systems*. Nonlinear systems frequently have more than one equilibrium point. Qualitative properties of nonlinear systems include the stability and instability of these equilibrium points in the sense of Lyapunov; boundedness of solutions; estimates of trajectory bounds; and input-output properties.

• *Limit Cycles*. Nonlinear systems can display fixed oscillations without external excitation. These oscillations are called limit cycles. Limit cycles represent an important phenomenon in nonlinear systems. They can be found in many areas of engineering and nature. An engineer has to know how to eliminate them when they are undesirable.

• *Bifurcations and Chaos*. In stable linear systems, small differences in parameters and in initial conditions can only cause small differences in the qualitative behavior. Nonlinear systems, however, can display dramatic changes in behavior caused by small changes in parameters and initial conditions. These changes are addressed in the theory of bifurcations which studies qualitative changes of system properties due to quantitative changes of parameters, and in the theory of chaos which studies systems which are extremely sensitive to initial conditions.

It is well known that the qualitative properties of *general nonlinear dynamical systems* are difficult to analyze. In this monograph, we will investigate the qualitative properties of a *specific class* of nonlinear systems–*dynamical systems with saturation type nonlinearities*. The topics addressed herein include *stability analysis, controllability, design methods*, as well as several *important applications*. In particular, we will investigate the *asymptotic stability* and the *null controllability* of a class of control systems with state saturation and control constraints, we will establish results for the *stability analysis* of one-dimensional and multidimensional state-space digital filters endowed with overflow nonlinearities, and we will develop a systematic *analysis and synthesis* procedure for a class of feedback neural networks with linear saturation activation functions and with arbitrarily prespecified interconnecting structures.

Under the assumption that a system operates in a *small* range (usually about the origin), linear models are used extensively to describe the dynamics of a system. However, when a system is required to operate over a *large* range, in most cases, linear models do not provide satisfactory descriptions of the system's behavior. Frequently, linear models are modified by imposing some types of nonlinearities which describe the dynamics outside the system's linear range. One of the frequently used nonlinearities is the *saturation* nonlinearity. Another assumption often used in linear system theory is that the system model is linearizable. However, in control systems there are many nonlinearities whose discontinuous nature does not allow linear approximation. In these cases, no linear models can be used to describe the dynamics of the system. Specific cases include *hard limiter* type nonlinearities, such as relay, dead-zone, backslash, and *saturation*. In the present research, we will investigate exclusively systems with saturation type nonlinearities.

Saturation nonlinearities arise very often in the modeling process of dynamical systems. (Examples of such systems are common in engineering and include mechanical systems with position and speed limits, electric systems with limited power supply for the actuators (motors), digital filters implemented in finite wordlength format, and so on.) In such cases, there are no linear models to describe the system's dynamics and saturation nonlinearities are *intrinsic* to the system. Qualitative analysis, especially stability analysis is a fundamental issue in the study of dynamical systems and therefore it is necessary to develop a theory for the qualitative analysis of general dynamical systems with saturation type nonlinearities, which include the above examples as special cases.

Generally, we will have in mind stability in the sense of Lyapunov, and frequently we will assume that a system has only one equilibrium, usually at the origin. Conditions under which a system with saturation type nonlinearities has a unique equilibrium at the origin and under which this equilibrium is *globally asymptotically stable* will be part of our primary interests. It is worth mentioning that the stability analysis of digital filters with saturation type nonlinearities (overflow nonlinearities) has been addressed for about three decades.

In the present research, we also consider systems which have more than on asymptotically stable equilibrium and in which the saturation nonlinearities are either intrinsic or are introduced intentionally. Feedback neural networks are examples of systems which have multiple asymptotically stable equilibrium points as memory points and in which the saturation nonlinearities are either intrinsic because of the nonlinear behavior of the components in the network, or are intentionally added in order to achieve certain desired properties (e.g., memory capabilities).

Feedback type artificial neural networks are a class of nonlinear dynamical systems. The qualitative analysis and design of such systems have been actively pursued research topics in the past ten years. Networks of this type consist of a number of basic (nonlinear) elements, usually called *neurons*, with weighted connections among these elements. Such networks are usually required to have many asymptotically stable equilibria as memory points. Our objective will be to establish results which enable us to determine all the equilibrium points and to ascertain the qualitative properties of these equilibrium points for a given neural network, and furthermore, to develop synthesis procedures which enable us to design a neural network so that the network has a given set of desired patterns as part of its asymptotically stable equilibria. A systematic *analysis and synthesis* procedure to achieve these objectives will be one of our final goals.

In an artificial neural network, the nonlinear elements, namely, the neurons, are key components. The input to each neuron consists of a weighted sum of neuron outputs. The nonlinearities used to model the neurons are usually nondecreasing functions called *activation functions* with values constrained to a closed interval (e.g., $[-1, 1]$). Among the various functions used to model neurons are the members of a class of piecewise linear functions–the linear *saturation* functions–which have received special attention. The qualitative analysis and design of artificial neural networks with linear saturation activation functions have received a great deal of attention and have achieved high levels of sophistication and maturity.

1.2   Control Systems with Saturation Nonlinearities

Control systems are excellent examples of dynamical systems. In the *first part* of this monograph, we consider a class of control systems in which the states of the system are endowed with saturation nonlinearities and the controller of the system is constrained to a compact set. Two fundamental issues will be considered for control systems of this type. Specifically, for the zero-input case, we investigate the *global asymptotic stability* of the null solution, and for the case with constrained controls, we investigate the *null controllability*.

As mentioned in the previous section, one sometimes assumes that a system functions in a small range of its operating point and that the system is described by a set of linear (differential or difference) equations. This is not always the case in practice where on many occasions we have to consider a large working range (in the state space) for a control system, and where the state of the system *saturates* when it reaches the physical limits of the system which are usually finite. Existing linear control theory suffers from the fundamental shortcoming that, in reality, dynamical systems are frequently subject to several complicating nonlinear effects which may invalidate or at least severely limit its applicability. Such nonlinearities are frequently typified by signal saturation, *i.e.*, the output (or states) of physical systems are usually limited to some degree and when a limiting value is reached, saturation is said to occur. Thus, examples of control systems in which the states are endowed with saturation nonlinearities abound.

For any given nonlinear system, the first fundamental question concerning its qualitative properties is under what conditions the system is *stable* (in the sense of Lyapunov), *i.e.*, under what conditions the system has a unique equilibrium at the origin and under what conditions this equilibrium is (globally) asymptotically stable. In the case of control systems, we usually concern ourselves with stability properties under the influence of initial conditions with no external inputs. In the present monograph, we establish new results for the stability analysis of both continuous-time and discrete-time control systems with state saturation and with zero-input–a special class of nonlinear dynamical systems. The stability results which we establish herein constitute original contributions.

Controllability is of fundamental interest in all control systems. In the case of nonlinear control systems, we usually consider *null controllability*, *i.e.*, the ability of the state of a (discrete-time) control system to be steered to the origin in a finite number of steps. For nonlinear control systems, in most cases one can only determine results for *local* null controllability; while in our case, which is a special case of nonlinear control systems–control systems with state saturation, we are able to determine *global* results for null controllability. Of particular interest is the null controllability of such systems when the control signal is constrained to some compact set. The null controllability of control systems with state saturation *and* control constraints is investigated for the first time in the present work. We establish some results on this topic, with further results being expected.

1.3 Digital Filters with Overflow Corrections

Digital filters implemented in finite wordlength format are typical nonlinear dynamical systems. Filters of this type are usually called fixed-point digital filters. In such systems, the nonlinearities are intrinsic and include *overflow nonlinearities* and quantization effects. Overflow nonlinearities correspond to the type of arithmetic that is used to cope with the overflow resulting from computation. The usual types of overflow arithmetic include *saturation, zeroing, triangular, two's complements*, and other characteristics. In the *second part* of this monograph, we establish *stability* results for digital filters endowed with overflow nonlinearities. Quantization effects will not be considered in the present research.

Overflow nonlinearities are usually represented by *piecewise linear functions* which include the saturation nonlinearity as a special case. We will consider a generalized overflow nonlinearity, which includes the usual types of overflow arithmetic used in practice as special cases. Our presentation in the second part of this volume is focused on systems with overflow nonlinearities rather than just systems with saturation nonlinearities, partially because the methodology developed in the first part of this monograph, which is for systems with saturation nonlinearities, is readily applicable to systems with overflow nonlinearities.

One of the fundamental requirements for fixed-point digital filters is that they be free of *limit cycles* (*overflow oscillations*) under zero-input. Limit cycles in fixed-point digital filters will result in unacceptable performance in most cases, and as far as we are concerned, limit cycles in fixed-point digital filters (with no external inputs) constitute undesirable phenomena. Presently, investigators are still investigating conditions to guarantee the non-existence of limit cycles in fixed-point digital filters which are not overly conservative. It is well-known that when the null solution of a digital filter is globally asymptotically stable, the filter will be free of limit cycles. This fact will be used in the development of our criteria for the non-existence of limit cycles in fixed-point digital filters.

The fixed-point digital filters mentioned above are typical one-dimensional, *time domain* nonlinear systems. In signal processing, we also frequently encounter multidimensional digital filters, which operate in *spatial domain* and are also realized in finite wordlength format. As in the one-dimensional case, the absence of limit cycles in multidimensional digital filters with overflow nonlinearities is also a fundamental requirement, and the global asymptotic stability of a multidimensional filter's null solution implies the non-existence of limit cycles in such filters.

We will establish our stability results for one-dimensional and multidimensional digital filters using Lyapunov's Second Method. Our emphasis will be on results involving quadratic form Lyapunov functions, and more importantly, we will establish necessary and sufficient conditions for generating such Lyapunov functions. Some of the existing results in the literature will also be reviewed. Results established herein are general, because we consider generalized overflow nonlinearities which are a generalization of the usual types of overflow arithmetic used in practice and because we establish necessary and sufficient conditions for generating quadratic form Lyapunov functions. The results of this part of the present monograph greatly improve existing results.

1.4 Feedback Neural Networks and Associative Memories

Feedback neural networks are important classes of nonlinear dynamical systems which have a multitude of asymptotically stable equilibrium points that serve as desired memories. In the *third part* of this monograph, we consider a class of feedback neural networks, which is a variant of the analog Hopfield model. In the original Hopfield model, one uses (continuously differentiable) sigmoidal functions as activation functions while in the present case, we employ the linear saturation function as activation functions.

Neural networks are good candidates for novel information processing systems. It is believed that these systems are generally robust with respect to malfunctions of individual devices because of the *distributed* nature of the flow of information. This is in contrast to digital computers, which heavily rely on the perfect functioning of each single device. Neural networks are highly interconnected systems of simple processors (neurons) which manipulate data in a *parallel* manner. This is again in contrast to most digital computers which operate serially and which contain some complex components. These attributes enable neural networks to solve complex problems, including pattern recognition, optimization, and *associative memories*, tasks which are sometimes difficult for digital computers to cope with.

Associative memories implemented by artificial feedback neural networks are composed of a collection of interconnected processing elements having features of high degree of parallelism and distributed storage of information. The information processing capabilities are partly due to the dynamical behavior of stable states of such networks which act as basins of attraction towards which neighboring states evolved in time. This time evolution of an array of neuron-like elements towards these equilibrium points can be interpreted as the evolution of an imperfect pattern moving towards a correct pattern. In other words, the process of association and information retrieval of associative memories are simulated by the dynamical behavior of a highly interconnected system of nonlinear circuit elements.

We are particularly interested in the implementations of associative memories (or, content addressable memories) via artificial neural networks. The goal of associative memories is to store a set of desired patterns in such a manner that the network will recognize a stored pattern if the input pattern contains a sufficiently large portion of that pattern's information. In our neural network model, the input pattern is the initial state of the neural network. In our applications of neural networks to associative memories, we will store desired patterns as asymptotically stable equilibrium points (or output vectors corresponding to a set of asymptotically stable equilibrium points) of the network, which implies that there will be a *domain of attraction* associated with each desired pattern, and if the initial state of the network is in the domain of attraction of a desired pattern, the network will eventually converge to that pattern.

Hopfield networks are usually fully interconnected and have been shown to effectively implement associative memories. Besides the Hopfield model, many other feedback neural network models, which are also fully interconnected, have been proposed for the implementation of associative memories (we will review further some of these neural network models in Chapter 8). Our primary interest here is the implementation of associative memories by means of neural networks with *partial or sparse* interconnections. *Cellular neural networks*, which are described by the same differential equations as those that are used for the Hopfield model and which have only local interconnections, are special cases of the sparsely interconnected neural networks. Existing analysis and synthesis results for neural networks with applications to associative memories are developed exclusively for fully interconnected cases. Applications of sparsely interconnected neural networks for associative memories, including cellular neural networks, have not successfully been used prior to the present work.

Our motivation for conducting this part of our research is due to serious problems encountered in the analog VLSI implementations of feedback neural networks. Large numbers of interconnections and extremely large numbers of line-crossings resulting from interconnections in fully interconnected neural networks make the analog VLSI implementations of such networks nearly impossible. Therefore, it is very desirable to develop *synthesis procedures for associative memories* via artificial neural networks with less interconnections and with less line-crossings among the interconnections. In particular, we will require that the designed neural networks have *arbitrarily prespecified* sparse interconnecting structures. Another problem which is also of great concern in the implementation of neural networks is that the usual implementation process in neural nets is not perfect in the sense that one cannot realize perfectly the computed parameters. Therefore, it is desirable to establish *robustness analysis* results which can be used to determine conditions under which a perturbed neural network has the same set of desired memories as the network without perturbations. We provide answers to the above two practical problems in the third part of this monograph. We believe that the contributions given in the third part of this work are of theoretical as well as practical importance.

PART I

QUALITATIVE THEORY OF CONTROL SYSTEMS WITH CONTROL CONSTRAINTS AND STATE SATURATION: TWO FUNDAMENTAL ISSUES

CHAPTER 2

INTRODUCTION TO PART I

2.1 Concepts of Stability and Lyapunov Functions

We introduce in this section necessary notation and definitions from the fundamental Lyapunov stability theory [88]. We begin by providing some of the notation which we shall require throughout.

If $x \in R^n$, then $\|x\|$ will denote norm of x, where $\|\cdot\|$ represents any one of the equivalent norms on R^n. Also, if A is any real $m \times n$ matrix, then $\|A\|$ will denote the norm of the matrix A induced by the norm on R^n, $i.e.$,

$$\|A\| = \sup_{\|x\|=1} \|Ax\| = \sup_{0 < \|x\| \le 1} \frac{\|Ax\|}{\|x\|} = \sup_{x \ne 0} \frac{\|Ax\|}{\|x\|}.$$

Note in particular that [85]

$$\|Ax\| \le \|A\| \, \|x\|.$$

We concern ourselves with discrete-time systems described by equations of the form

$$x(k+1) = f(x(k)), \ k = 0, 1, 2, \cdots \tag{2.1}$$

and with continuous-time systems described by equations of the form

$$dx(t)/dt = f(x(t)), \ t \ge 0 \tag{2.2}$$

where $x \in R^n$. When discussing global results, such as global asymptotic stability, we shall always assume that $f: R^n \to R^n$. On the other hand, when considering local results, we shall usually assume that $f: B(h) \to R^n$ for some $h > 0$, where $B(h) \triangleq \{x \in R^n: \|x\| < h\}$. In either case, we will usually assume that the function $f(x)$ is continuous in x.

Definition 2.1 A point $x_e \in R^n$ is called an *equilibrium point* of (2.1) if $x_e = f(x_e)$. ∎

Definition 2.2 A point $x_e \in R^n$ is called an *equilibrium point* of (2.2) if $f(x_e) = 0$. ∎

Other terms for equilibrium point include *stationary point, singular point, critical point,* and *rest point.*

Definition 2.3 An equilibrium point x_e of (2.1) [resp., (2.2)] is called an *isolated equilibrium point* if there is an $h > 0$ such that $B(x_e, h) \triangleq \{x \in R^n : \|x - x_e\| < h\}$ contains no equilibrium point of (2.1) [resp., (2.2)] other than x_e itself. ∎

We now give precise definitions of several stability concepts. We assume that (2.1) [resp., (2.2)] possesses an isolated equilibrium at the origin. Thus, $f(0) = 0$.

Definition 2.4 The equilibrium $x_e = 0$ of (2.1) [resp., (2.2)] is *stable* if for every $\varepsilon > 0$ there exists a $\delta(\varepsilon) > 0$ such that $\|x(k)\| < \varepsilon$ for all $k \geq 0$ whenever $\|x(0)\| < \delta(\varepsilon)$ [resp., $\|x(t)\| < \varepsilon$ for all $t \geq 0$ whenever $\|x(0)\| < \delta(\varepsilon)$]. ∎

Definition 2.5 The equilibrium $x_e = 0$ of (2.1) [resp., (2.2)] is *asymptotically stable* if

(*i*) it is stable, and

(*ii*) there exists an $\eta > 0$ such that $\lim_{k \to \infty} x(k) = 0$ whenever $\|x(0)\| < \eta$ [resp., $\lim_{t \to \infty} x(t) = 0$ whenever $\|x(0)\| < \eta$]. ∎

The set of all $x(0) \in R^n$ such that $x(k) \to 0$ as $k \to \infty$ [resp., $x(t) \to 0$ as $t \to \infty$] is called the *domain of attraction* of the equilibrium $x_e = 0$ of (2.1) [resp., (2.2)] . Also, if for (2.1) [resp., (2.2)] condition (*ii*) in Definition 2.5 is true, then the equilibrium $x_e = 0$ is said to be *attractive.*

Definition 2.6 The equilibrium $x_e = 0$ of (2.1) [resp., (2.2)] is *asymptotically stable in the large* (or *globally asymptotically stable*) if it is stable and if every solution of (2.1) tends to zero as $k \to \infty$ [resp., every solution of (2.2) tends to zero as $t \to \infty$]. ∎

In the case of Definition 2.6, the domain of attraction of the equilibrium $x_e = 0$ of (2.1) [resp., (2.2)] is all of R^n. Note that in this case, $x_e = 0$ is the *only* equilibrium of (2.1) [resp., (2.2)] .

The concepts introduced in Definitions 2.4–2.6 are usually referred to as stability *in the sense of Lyapunov.*

The stability results of the equilibrium $x_e = 0$ of system (2.1) involve the existence of real valued functions $v: D \to R$. In the case of local results (e.g., stability and asymptotic stability results), we shall usually only require that $D = B(h) \subset R^n$ for some $h > 0$. On the other hand, in the case of global results (asymptotic stability in the large), we have to assume that $D = R^n$. Unless stated otherwise, we shall always assume that $v(0) = 0$.

If v is continuous with respect to all of its arguments, then we obtain the first forward difference of v with respect to k along the solutions of (2.1) (or along the trajectories of (2.1)), $Dv_{(2.1)}$, as

$$Dv_{(2.1)}(x(k)) = v(x(k+1)) - v(x(k)) = v(f(x(k))) - v(x(k)). \qquad (2.3)$$

It is important to note that in (2.3), the first forward difference of v with respect to k, along the solutions of (2.1), is evaluated *without having to solve equation (2.1).* The significance of this will become clear in the sequel.

If v is continuously differentiable with respect to all of its arguments, then we obtain the time derivative of v with respect to t, along the solutions of (2.2), $Dv_{(2.2)}$, as

$$Dv_{(2.2)} = \nabla v(x)^T f(x) \qquad (2.4)$$

where $\nabla v(x) = (\partial v/\partial x_1, \cdots, \partial v/\partial x_n)^T$.

We now give several important properties which v functions may possess.

Definition 2.7 A continuous function $v: R^n \to R$ [resp., $v: B(h) \to R$] is said to be *positive definite* if

(i) $v(0) = 0$, and

(ii) $v(x) > 0$ for $x \neq 0$ [resp., $0 < \|x\| \leq r$ for some $r > 0$]. ∎

Definition 2.8 A continuous function $v: R^n \to R$ is said to be *radially unbounded* if

(i) $v(0) = 0$,

(ii) $v(x) > 0$ for all $x \in R^n - \{0\}$, and

(iii) $v(x) \to \infty$ as $\|x\| \to \infty$. ∎

Definition 2.9 A function v is said to be *negative definite* if $-v$ is a positive definite function. ▪

Definition 2.10 A continuous function $v \colon R^n \to R$ [resp., $v \colon B(h) \to R$] is said to be *positive semidefinite* if

(*i*) $v(0) = 0$, and

(*ii*) $v(x) \geq 0$ for all $x \in B(r)$ for some $r > 0$. ▪

Definition 2.11 A function v is said to be *negative semidefinite* if $-v$ is positive semidefinite. ▪

Some of the preceding characterizations of v functions can be rephrased in equivalent and very useful ways. In doing so, we employ certain comparison functions which we introduce next.

Definition 2.12 A continuous function $\psi \colon [0, r_1] \to R^+$ [resp., $\psi \colon [0, \infty) \to R^+$] is said to belong to *class* $\mathcal{K}$, i.e., $\psi \in \mathcal{K}$, if $\psi(0) = 0$ and if ψ is strictly increasing on $[0, r_1]$ [resp., on $[0, \infty)$]. If $\psi \colon R^+ \to R^+$, if $\psi \in \mathcal{K}$, and if $\lim_{r \to \infty} \psi(r) = \infty$, then ψ is said to belong to *class* $\mathcal{KR}$. ▪

We are now in a position to state the following results. Their proofs can be found in [88].

Theorem 2.1 A continuous function $v \colon R^n \to R$ [resp., $v \colon B(h) \to R$] is *positive definite* if and only if

(*i*) $v(0) = 0$, and

(*ii*) for any $r > 0$ [resp., some $r > 0$] there exists a $\psi \in \mathcal{K}$ such that $v(x) \geq \psi(\|x\|)$ for all $x \in B(r)$. ▪

We remark that both of the equivalent definitions of positive definite functions just given will be used. One of these forms is often easier to use in specific examples, while the second form will be very useful in proving stability results.

Theorem 2.2 A continuous function $v\colon R^n \to R$ is *radially unbounded* if and only if

(i) $v(0) = 0$, and

(ii) there exists a $\psi \in \mathcal{KR}$ such that $v(x) \geq \psi(\|x\|)$ for all $x \in R^n$. ∎

We close the present section with a discussion of an important class of v functions. Let $x \in R^n$, let $B = [b_{ij}]$ be a real symmetric $n \times n$ matrix, and consider the *quadratic form* $v\colon R^n \to R$ given by

$$v(x) = x^T B x = \sum_{i,k=1}^{n} b_{ik} x_i x_k. \tag{2.5}$$

Recall that in this case, B is diagonalizable and all of its eigenvalues are real. The next theorem is due to Sylvester.

Theorem 2.3 Let v be the quadratic form defined in (2.5). Then

(i) v is positive definite (and radially unbounded) if and only if all principal minors of B are positive, *i.e.*, if and only if

$$\det \begin{bmatrix} b_{11} & \cdots & b_{1k} \\ \vdots & & \vdots \\ b_{k1} & \cdots & b_{kk} \end{bmatrix} > 0, \quad k = 1, \cdots, n.$$

(These inequalities are called the *Sylvester inequalities*.)

(ii) v is negative definite if and only if

$$(-1)^k \det \begin{bmatrix} b_{11} & \cdots & b_{1k} \\ \vdots & & \vdots \\ b_{k1} & \cdots & b_{kk} \end{bmatrix} > 0, \quad k = 1, \cdots, n.$$

∎

We will require the following definitions.

Definition 2.13 A symmetric matrix $B = [b_{ij}]$ is said to be *positive definite* if all the eigenvalues of B are positive. ∎

Definition 2.14 A symmetric matrix $B = [b_{ij}]$ is said to be *positive semidefinite* if all the non-zero eigenvalues of B are positive. ∎

The following results (cf. [85]) are very important in applications.

Theorem 2.4 Let v be the quadratic form defined in (2.5). Then

(i) v is positive definite if and only if B is a positive definite matrix.

(ii) v is positive semidefinite if and only if B is a positive semidefinite matrix. ∎

It is customary in the literature to write $B > 0$ when B is positive definite and to write $B \geq 0$ when B is positive semidefinite.

2.2 Principal Lyapunov Stability Theorems

In this section we present results from the Lyapunov theory which serve as essential background material for the first and second parts of this monograph. We give precise statements of several well-known stability results [88] for the system of equations given by (2.1) and (2.2). These results comprise the *direct method of Lyapunov* for discrete-time systems and continuous-time systems, which is also sometimes called the *Second Method of Lyapunov*. The reason for this nomenclature is clear: results of the type presented here allow us to make qualitative statements about whole families of solutions of (2.1) and (2.2) without actually solving these equations.

As already mentioned, in the case of local stability results, we shall require that $x_e = 0$ is an isolated equilibrium of (2.1) or of (2.2), and in the case of global stability results, we shall require that $x_e = 0$ is the only equilibrium of (2.1) or of (2.2).

The results given in the following require the existence of functions $v\colon B(h) \to R$ (resp., $v\colon R^n \to R$) with the indicated properties. In these results, we concern ourselves with the stability, asymptotic stability, and global asymptotic stability of the equilibrium $x_e = 0$ of (2.1) and (2.2). For other stability results, please refer to [88]. The proofs of these results can be found in [88].

Theorem 2.5 If there exists a continuous, positive definite function v with a negative semidefinite (or identically zero) first forward difference $Dv_{(2.1)}$ (see (2.3)), then the equilibrium $x_e = 0$ of (2.1) is *stable*. ∎

Theorem 2.6 If there exists a continuous, positive definite function v with a negative definite first forward difference $Dv_{(2.1)}$, then the equilibrium $x_e = 0$ of (2.1) is *asymptotically stable*. ∎

Theorem 2.7 If there exists a continuous, positive definite, and radially unbounded function v such that $Dv_{(2.1)}$ is negative definite for all $x \in R^n$, then the equilibrium $x_e = 0$ of (2.1) is *asymptotically stable in the large*. ∎

Theorem 2.8 If there exists a continuously differentiable, positive definite function v such that $Dv_{(2.2)}$ is identically zero or negative semidefinite (see (2.4)), then the equilibrium $x_e = 0$ of (2.2) is *stable*. ∎

Theorem 2.9 If there exists a continuously differentiable, positive definite function v with negative definite $Dv_{(2.2)}$, then the equilibrium $x_e = 0$ of (2.2) is *asymptotically stable*. ∎

Theorem 2.10 If there exists a continuously differentiable, positive definite, and radially unbounded function v such that $Dv_{(2.2)}$ is negative definite for all $x \in R^n$, then the equilibrium $x_e = 0$ of (2.2) is *asymptotically stable in the large*. ∎

Results of the type given above are referred to as *Lyapunov stability theorems*. Any function v which satisfies any one of the above results will be called a *Lyapunov function*. We will use these theorems in Chapters 3, 6, and 7.

The Lyapunov theorems are very powerful. However, in general, difficulties may arise in applying these results, especially to high-dimensional (*i.e.*, high-order) systems with complicated structure. The reason for this lies in the fact that there is no universal and systematic procedure available which tells us how to find appropriate Lyapunov functions. For this reason, we will focus on several special forms of Lyapunov functions, including l_p vector norms and quadratic forms.

2.3 Controllability

Some of the most significant results in linear systems concern Kalman's controllability and observability theory [48]. Controllability and observability are properties that describe structural features of a dynamical system. A (discrete-time) system is said to be controllable if for every x_0 and x_1 there is a finite N and a sequence of controls $u(0), \cdots, u(N)$ such that if the system has state x_0 at time $k = 0$, it is forced to state x_1 at $k = N$. In Kalman's results, there are no constraints on either states or controls of the system. In practice, it is frequently desirable to use nonlinear and/or constrained control laws to achieve a system's full capability [35], [43]. Also, there are practical considerations for using constrained controls, since the energy of control signals is usually limited.

E. B. Lee and L. Markus [55] were among the first to consider the controllability of linear systems with constrained controls. Their motivation was the fact that the notion of controllability of linear systems with constrained controls is very important for certain optimal control problems, including minimum energy control problem. They considered linear systems described by

$$\dot{x}(t) = Ax(t) + Bu(t), \tag{2.6}$$

where $A \in R^{n \times n}$, $B \in R^{n \times m}$, $x \in R^n$, $u \in \Omega \subset R^m$ and the control constraint set Ω is convex and compact and contains the origin in its interior. System (2.6) is said to be *globally null controllable* if, given any initial state $x(0) \in R^n$, the state of the system can be steered to the origin by using $u \in \Omega$ in a finite time interval. Lee and Markus showed in their book (see, [55], page 92) that necessary and sufficient conditions for system (2.6) to be globally null controllable, using constrained control $u \in \Omega$, are

(i) $\text{rank}[B, AB, \cdots, A^{n-1}B] = n$, *i.e.*, (A, B) is controllable; and

(ii) every eigenvalue λ of A satisfies $Re\lambda(A) \leq 0$.

Our current research on controllability will be focused on discrete-time systems. Frequently, a theory for discrete-time dynamical systems can be developed, side by side, with counterpart results for continuous-time dynamical systems. The reason we develop our results for discrete-time systems separately is that our progress on the stability analysis for discrete-time systems has been greater than that for continuous-time systems and since we will make use of these stability results in the investigation of the controllability problem.

The discrete-time counterpart of the above result was given by E. D. Sontag [109]. In the discrete-time case, the system has the form

$$x(k + 1) = Ax(k) + Bu(k), \tag{2.7}$$

where A, B, x, and u are defined similarly as in (2.6). System (2.7) is said to be globally null controllable if its state can be steered to the origin by using $u(k) \in \Omega$ in a finite number of steps for any given initial state $x(0) \in R^n$. We will review some of the results for discrete-time systems in Chapter 4.

Control systems with saturation on controllers are still under investigation (see e.g., [15], [16], [17], [37], [49], [53], [54]). These studies are concerned with the stabilization of linear dynamical systems with saturated controllers. Also, in these studies, it is generally assumed that there is no state saturation in the system. In practice, this is not realistic. For example, in describing the dynamics of a motor vehicle, we may choose speed and steering angle as two of the state variables. Since both of these variables have upper and lower limits, this system is encumbered with state saturation nonlinearities. In applications, state saturation in control systems is very common.

For systems with state saturation and control constraints, there are many open problems that need investigating. One of the open problems is *null controllability*. Controllability is a fundamental issue in control systems, and for nonlinear control systems is difficult to investigate. Our attempts at solving the null controllability problem of systems with state saturation and control constraints have been fruitful thus far (cf. [60], [61], [63]–[73], or Chapters 3, 6, 7, 9–11, and [62] or Chapter 4).

CHAPTER 3

ASYMPTOTIC STABILITY OF DYNAMICAL SYSTEMS WITH STATE SATURATION

3.1 Introduction to Continuous-Time Systems

In this chapter, we first investigate stability properties of continuous-time systems described by

$$\dot{x}(t) = h[Ax(t)], \quad t \geq 0 \tag{3.1}$$

where $x(t) \in D^n \triangleq \{x \in R^n : -1 \leq x_i \leq 1, \; i = 1, \cdots, n\}$, $A = [a_{ij}] \in R^{n \times n}$,

$$h(Ax) = \left[h\left(\sum_{j=1}^{n} a_{1j}x_j \right), \cdots, h\left(\sum_{j=1}^{n} a_{nj}x_j \right) \right]^T,$$

and

$$h\left(\sum_{j=1}^{n} a_{ij}x_j \right) = \begin{cases} 0, & x_i = 1 \\ \sum_{j=1}^{n} a_{ij}x_j, & -1 < x_i < 1 \\ 0, & x_i = -1 \end{cases}, \quad \text{for } i = 1, \cdots, n.$$

We will say that *system (3.1) is stable* if $x_e = 0$ is the globally asymptotically stable equilibrium, *i.e.*, if $x_e = 0$ is stable and if $x(t) \to 0$ as $t \to \infty$ for any $x(0) \in D^n$. We will refer to (3.1) as a *"linear" system operating on a closed hypercube*. (D^n represents the closed unit hypercube.)

Equation (3.1) represents a class of continuous-time dynamical systems with symmetrically saturating states after normalization. Examples of such systems include control systems [60] and certain classes of neural networks [59], [87].

When considering (3.1) as a *control system* (with no external inputs), some of the first fundamental questions that arise concern the existence and uniqueness of an equilibrium or operating point (which we assume to be the origin, without loss of generality [88]) and the qualitative properties (specifically, stability properties) of such an equilibrium. The condition that the matrix A in system (3.1) be stable (*i.e.*, that all of the eigenvalues of A be located in the open left half complex plane) does not ensure that $x_e = 0$ is a unique equilibrium, and hence, it does not ensure that $x_e = 0$ is asymptotically stable in the large. For example, the matrix

$$A = \begin{bmatrix} 11.1 & -20 & 4 & -7 \\ 30 & -30 & -1 & -19.5 \\ 8.4 & 6.6 & 10 & -20 \\ 10 & -10 & 30 & -30 \end{bmatrix}$$

has eigenvalues

$$\lambda(A) = -0.2921, \ -28.5009, \ -5.0535 \pm 21.6362i,$$

i.e., A is stable. It is easily verified that in addition to the origin, the system (3.1) with A specified above, has also equilibria at

$$x_a = [-1, \ -0.3167, \ -1, \ -1]^T,$$

and

$$x_b = [1, \ 0.3167, \ 1, \ 1]^T.$$

Thus, while $x_e = 0$ is certainly asymptotically stable, it is not asymptotically stable in the large.

When considering system (3.1) as a *neural network with applications to optimization problems,* we wish to construct a network with a unique equilibrium which is globally asymptotically stable, in order to prevent convergence to local minima of an objective function (see, e.g., [46], [112]). When the desired equilibrium x_d is located in the interior of D^n, the conditions for this equilibrium x_d to be globally asymptotically stable will be identical to the conditions for the equilibrium $x_e = 0$ of (3.1) to be globally asymptotically stable, since we can always consider $x_d = 0$, without loss of generality (cf. [88]).

In the first part of the present chapter (Sections 3.1–3.3), we will establish a sufficient condition which ensures the global asymptotic stability of the equilibrium $x_e = 0$ of system (3.1).

In Section 3.2, we introduce some essential notation. In Section 3.3, we present our main result for the the stability of continuous-time system (3.1). A few additional pertinent remarks can be found in Section 3.8.

3.2   Notation

Before proceeding further, we introduce the following notation.

For $x \in R^n$, we define the l_p vector norm as

$$\|x\|_p = (\sum_{i=1}^{n} |x_i|^p)^{1/p}, \quad \text{for } 1 \leq p \leq \infty.$$

Recall that when $p = \infty$, we have

$$\|x\|_\infty = \max_{1 \leq i \leq n} \{|x_i|\}.$$

For $A \in R^{n \times n}$, we define the norm of A by

$$\|A\| = \inf\{\gamma : \|Ax\| \leq \gamma\|x\| \text{ for all } x \in R^n\}.$$

Recall that for $p = \infty$, the norm of A, induced by the l_∞ vector norm, is given by

$$\|A\|_\infty = \max_{1 \leq i \leq n} \{\sum_{j=1}^{n} |a_{ij}|\}.$$

The measure of a matrix $A \in R^{n \times n}$ is defined as

$$\mu_p(A) = \lim_{\theta \to 0+} \frac{\|I + \theta A\|_p - 1}{\theta}, \tag{3.2}$$

where $\|\cdot\|_p$ denotes the matrix norm induced by the l_p vector norm and I is the identity matrix. In particular, when $p = \infty$, we have

$$\mu_\infty(A) = \max_{1 \leq i \leq n} \{a_{ii} + \sum_{j=1, j \neq i}^{n} |a_{ij}|\}.$$

We denote the interior and the boundary of a set Ω by $(\Omega)^0$ and $\partial\Omega$, respectively.

For $x = [x_1, \cdots, x_n]^T$ and $y = [y_1, \cdots, y_n]^T$, we let

$$x * y = [x_1 y_1, \cdots, x_n y_n]^T,$$

and we let

$$\min(x) = \min_{1 \leq i \leq n} \{x_i\}.$$

Also, the notation $x \leq y$ will mean $x_i \leq y_i$ for $1 \leq i \leq n$.

If $A = [A_{ij}]$ is an arbitrary matrix, then A^T denotes the transpose of A. If A is a square matrix, we use $\lambda(A)$ to denote the set of eigenvalues of matrix A.

Let P(n) denote the set of all permutations on $\{1, \cdots, n\}$.

3.3 Main Result for Continuous-Time Systems

We recall that for a general autonomous system

$$\dot{x} = f(x), \tag{3.3}$$

with $x \in R^n$ and $f : R^n \to R^n$, x_e is an equilibrium for (3.3) if and only if

$$f(x_e) = 0.$$

We can assume, without loss of generality, that $x_e = 0$ (see, e.g., [88]). Thus, for system (3.1), we assume that A is nonsingular.

We are now in a position to establish the following result.

Theorem 3.1 The equilibrium $x_e = 0$ of system (3.1) is globally asymptotically stable, if

$$\mu_\infty(A) < 0. \tag{3.4}$$

Since $Re\lambda(A) \leq \mu_\infty(A) < 0$ (see, [28]), the equilibrium $x_e = 0$ is clearly asymptotically stable. We need only to prove that it is also *globally* asymptotically stable. We will prove this in the following two steps.

(1) For any $x(0) \in D^n$, $x(t)$ will not always stay on ∂D^n for $t > 0$.

(2) $x(t) \to 0$ as $t \to \infty$ for any $x(0) \in (D^n)^0$.

In the proof of the first step, we will utilize the notation given below, which was first introduced in [59] and [87].

For each integer m, $0 \leq m \leq n$, let

$$\Lambda_m = \left\{ \xi = [\xi_1, \cdots, \xi_n]^T \in \Lambda : \xi_{\sigma(i)} = 0, \ 1 \leq i \leq m \ \text{and} \ \xi_{\sigma(i)} = \pm 1, \ m < i \leq n, \right.$$

$$\left. \text{for some } \sigma \in P(n) \right\}$$

where

$$\Lambda = \{ \xi = [\xi_1, \cdots, \xi_n]^T : \xi_i = \pm 1 \ \text{or} \ 0, \ 1 \leq i \leq n \}.$$

For each $\xi \in \Lambda$, let

$$C(\xi) = \{x = [x_1, \cdots, x_n]^T \in R^n : |x_i| < 1 \text{ if } \xi_i = 0, \text{ and } x_i = \xi_i \text{ if } \xi_i \neq 0\}.$$

Suppose that $\xi \in \Lambda_m$ and $\sigma \in P(n)$ such that

$$\xi_{\sigma(i)} = 0, \ 1 \leq i \leq m, \text{ and } \xi_{\sigma(i)} = \pm 1, \ m < i \leq n. \tag{3.5}$$

We denote

$$A_{I,I} = [a_{\sigma(i)\sigma(j)}]_{1 \leq i,j \leq m},$$

$$A_{I,II} = [a_{\sigma(i)\sigma(j)}]_{1 \leq i \leq m, m < j \leq n},$$

$$A_{II,I} = [a_{\sigma(i)\sigma(j)}]_{m < i \leq n, 1 \leq j \leq m},$$

$$A_{II,II} = [a_{\sigma(i)\sigma(j)}]_{m < i,j \leq n},$$

$$\xi_I = [\xi_{\sigma(1)}, \cdots, \xi_{\sigma(m)}]^T,$$

and

$$\xi_{II} = [\xi_{\sigma(m+1)}, \cdots, \xi_{\sigma(n)}]^T.$$

Remark 3.1 For a given $\xi \in \Lambda_m$, there may exist different elements in $P(n)$ for which (3.5) is true. For these different permutations, the notation given above will be the same up to different orders in the components. Thus, the subsequent analysis and conclusions will be identical for any of the permutations used. ∎

Remark 3.2 if $m = n$, we have $A_{I,I} = A, \xi_I = \xi$ and the $A_{I,II}, A_{II,I}, A_{II,II}, \xi_{II}$ do not exist. If $m = 0$, we have $A_{II,II} = A, \xi_{II} = \xi$ and the $A_{I,I}, A_{I,II}, A_{II,I}, \xi_I$ do not exist. ∎

Proof of Theorem 3.1:

(1) Consider $\xi \in \Lambda_m$, $0 < m < n$, with $\sigma \in P(n)$ such that $\xi_{\sigma(i)} = 0$, $1 \leq i \leq m$, and $\xi_{\sigma(i)} = \pm 1$, $m < i \leq n$. When $x \in C(\xi)$, system (3.1) becomes

$$\begin{cases} \dot{x}_I = A_{I,I}x_I + A_{I,II}\xi_{II} \\ \dot{x}_{II} = 0 \end{cases}, \tag{3.6}$$

where

$$x_I = [x_{\sigma(1)}, \cdots, x_{\sigma(m)}]^T,$$

with $-1 < x_{\sigma(i)} < 1$ for $1 \le i \le m$, and

$$x_{II} = [x_{\sigma(m+1)}, \cdots, x_{\sigma(n)}]^T = [\xi_{\sigma(m+1)}, \cdots, \xi_{\sigma(n)}]^T = \xi_{II}.$$

In order to satisfy $\dot{x}_{II} = 0$, i.e., in order to maintain x in $C(\xi)$, it is necessary that (cf. [59])

$$\min\left((A_{II,I}x_I + A_{II,II}\xi_{II}) * \xi_{II}\right) \ge 0. \tag{3.7}$$

Denote $A_{II,I} = [a_{ij}^{(1)}] \in R^{(n-m)\times m}$, $A_{II,II} = [a_{ij}^{(2)}] \in R^{(n-m)\times(n-m)}$,

$$x_I = [x_{\sigma(1)}, \cdots, x_{\sigma(m)}]^T = [x_1^{(1)}, \cdots, x_m^{(1)}]^T,$$

and

$$\xi_{II} = [\xi_{\sigma(m+1)}, \cdots, \xi_{\sigma(n)}]^T = [\xi_1^{(2)}, \cdots, \xi_{n-m}^{(2)}]^T.$$

Then, we have

$$(A_{II,I}x_I + A_{II,II}\xi_{II}) * \xi_{II} = (A_{II,I}x_I) * \xi_{II} + (A_{II,II}\xi_{II}) * \xi_{II}$$

$$= \left[\xi_1^{(2)} \sum_{j=1}^m a_{1j}^{(1)} x_j^{(1)}, \cdots, \xi_{n-m}^{(2)} \sum_{j=1}^m a_{n-m,j}^{(1)} x_j^{(1)}\right]^T$$

$$+ \left[\xi_1^{(2)} \sum_{j=1}^{n-m} a_{1j}^{(2)} \xi_j^{(2)}, \cdots, \xi_{n-m}^{(2)} \sum_{j=1}^{n-m} a_{n-m,j}^{(2)} \xi_j^{(2)}\right]^T. \tag{3.8}$$

By noting that $|x_i^{(1)}| = |x_{\sigma(i)}| < 1$ for $0 \le i \le m$ and $\xi_i^{(2)} = \xi_{\sigma(i)} = \pm 1$ for $m < i \le n$, we have

$$(A_{II,I}x_I + A_{II,II}\xi_{II}) * \xi_{II} \le \left[\sum_{j=1}^m |a_{1j}^{(1)}|, \sum_{j=1}^m |a_{2j}^{(1)}|, \cdots, \sum_{j=1}^m |a_{n-m,j}^{(1)}|\right]^T +$$

$$\left[a_{11}^{(2)} + \sum_{j=2}^{n-m} |a_{1j}^{(2)}|, a_{22}^{(2)} + \sum_{j=1,j\ne2}^{n-m} |a_{2j}^{(2)}|, \cdots, a_{n-m,n-m}^{(2)} + \sum_{j=1}^{n-m-1} |a_{n-m,j}^{(2)}|\right]^T$$

$$= \left[a_{11}^{(2)} + \sum_{j=1}^m |a_{1j}^{(1)}| + \sum_{j=2}^{n-m} |a_{1j}^{(2)}|, a_{22}^{(2)} + \sum_{j=1}^m |a_{2j}^{(1)}| + \sum_{j=1,j\ne2}^{n-m} |a_{2j}^{(2)}|, \cdots,\right.$$

$$\left. a_{n-m,n-m}^{(2)} + \sum_{j=1}^m |a_{n-m,j}^{(1)}| + \sum_{j=1}^{n-m-1} |a_{n-m,j}^{(2)}|\right]^T. \tag{3.9}$$

Notice that the entries in the right hand side of (3.9) are just rearrangements of

$$a_{ii} + \sum_{j=1,j\ne i}^n |a_{ij}|, \quad \text{for } i = \sigma(m+1), \cdots, \sigma(n),$$

and thus, since $\mu_\infty(A) < 0$ (by assumption), every entry in the right hand side of (3.9) is less than 0. Therefore, condition (3.7) will fail to hold and this is true for every m, $0 < m < n$. It is also true for $m = 0$ by noting that $(A\eta) * \eta < 0$ for any $\eta \in \Lambda_0$ when $\mu_\infty(A) < 0$. Thus, the state of system (3.1) will not stay on the boundary of D^n for all time t, since

$$\partial D^n = \bigcup_{m=0}^{n-1} \{C(\xi) : \xi \in \Lambda_m\}.$$

(2) Thus far, we have proved that for any $x(0) \in D^n$, it is impossible for $x(t)$ to remain in ∂D^n for *all* $t > 0$. We now show that under the conditions of our theorem, once $x(t)$ leaves ∂D^n, it will never enter ∂D^n again.

Since $x(t)$ will not always stay on ∂D^n, without loss of generality, we assume that $x(0) \in (D^n)^0$. Then, system (3.1) is equivalent to

$$\dot{x} = Ax \tag{3.10}$$

as long as $x(t)$ does not reach ∂D^n and the solution for (3.10) is given by

$$x(t) = e^{At}x(0).$$

We compute

$$\|x(t)\|_\infty = \|e^{At}x(0)\|_\infty \leq \|e^{At}\|_\infty \|x(0)\|_\infty.$$

Using the fact that (see [28], page 59)

$$\|e^{At}\|_p \leq e^{\mu_p(A)t}, \quad \text{for any } t \geq 0 \text{ and any } p \geq 1,$$

we have

$$\|x(t)\|_\infty \leq e^{\mu_\infty(A)t}\|x(0)\|_\infty < \|x(0)\|_\infty, \quad \text{for any } t > 0,$$

since $\mu_\infty(A) < 0$. This in turn implies that

$$x(t) \in (D^n)^0, \quad \text{for all } t > 0,$$

i.e., when $x(0) \in (D^n)^0$, $x(t)$ will *never* reach the boundary of D^n. Therefore, system (3.1) is equivalent to (3.10) *for all* $t \geq 0$ when $x(0) \in (D^n)^0$. Hence, $x(t) \to 0$ as $t \to \infty$, since $Re\lambda(A) < 0$. ∎

Summarizing, above we have shown that

(i) $x(0) \notin D^n$ is not allowed;

(ii) if $x(t_1) \in \partial D^n$, $x(t)$ cannot stay in ∂D^n for all $t > t_1$; and

(iii) once $x(t)$ is in $(D^n)^0$, it will never enter ∂D^n and $x(t) \to 0$ as $t \to \infty$.

Remark 3.3 As will be seen, the result in Theorem 3.1 constitutes a (somewhat conservative) continuous-time counterpart for a stability result for the discrete-time case to be introduced in the next section. More general results for the stability of system (3.1) which are less conservative are still under investigation. ∎

3.4 Introduction to Discrete-Time Systems

In the remainder of this chapter, we will investigate stability properties of discrete-time systems described by

$$x(k+1) = \text{sat}[Ax(k)], \quad k = 0, 1, 2, \cdots \tag{3.11}$$

where $x(k) \in D^n$, $A \in R^{n \times n}$,

$$\text{sat}(x) = [\text{sat}(x_1), \text{sat}(x_2), \cdots, \text{sat}(x_n)]^T,$$

and

$$\text{sat}(x_i) = \begin{cases} 1, & x_i > 1 \\ x_i, & -1 \le x_i \le 1 \\ -1, & x_i < -1 \end{cases}.$$

We will say that *system (3.11) is stable* if $x_e = 0$ is the only equilibrium of system (3.11) and $x_e = 0$ is globally asymptotically stable. (See Section 2.1.) Also, since we have saturation nonlinearities in (3.11), it is clear that for any $x(0) \notin D^n$, $x(k) \in D^n$, $k \ge 1$, will always be true. Thus, without loss of generality, we will assume that $x(0) \in D^n$.

System (3.11) is the discrete-time counterpart of system (3.1). Equation (3.11) describes a class of discrete-time dynamical systems with symmetrically saturating states after normalization. Examples of such systems include control systems having saturation type nonlinearities on the state (cf. [36], [61], [68], [89]); neural networks defined on hypercubes (cf. [59], [87]); and digital filters using saturation overflow arithmetic (see, e.g., [6], [7], [8], [11], [12], [24], [25], [31], [33], [51], [61], [67], [90], [91], [92], [98], [102], [103], [106], [107], [108], [114], [120]).

1. System (3.11) may be used to represent *control systems* with saturating states with

no external inputs. In the analysis and design of such systems, the first fundamental question addresses stability: under what conditions is $x_e = 0$ an equilibrium and when is this equilibrium globally asymptotically stable?

The condition that A is a stable matrix, *i.e.*, every eigenvalue λ_i of A satisfies $|\lambda_i| < 1$, is not sufficient for system (3.11) to be stable. (It is easy to give examples for which A is a stable matrix, but system (3.11) is not stable). One way of guaranteeing the stability of system (3.11) is to consider D^n as a state constraint set which is positively invariant and contractive [9], [10], [116] with respect to the linear system

$$x(k+1) = Ax(k) \tag{3.12}$$

(*i.e.*, for (3.12), $x \in D^n$ implies $Ax \in D^n$, and if $x(0) \in D^n$, then $x(k) \to 0$, as $k \to \infty$). This is true *if and only if*

$$\|A\|_\infty < 1, \tag{3.13}$$

where $\| \cdot \|_\infty$ represents the matrix norm induced by the l_∞ vector norm. Condition (3.13) will guarantee the global asymptotic stability of the equilibrium $x_e = 0$ of system (3.11) since under this condition, system (3.11) and system (3.12) are equivalent. Note that condition (3.13) is the discrete-time counterpart of condition (3.4).

Condition (3.13) may also be viewed as a direct consequence of the results in [9], [10], [116], where necessary and sufficient conditions for a polyhedral state constraint set to be positively invariant and contractive are given. We point here to the difference between a system with state saturation nonlinearity and a system with state constraint set D^n. The former is a system with a nonlinear property, while the latter is a system whose states are not allowed to violate a constraint set. It is expected that the condition that system (3.11) is stable should be less conservative than the condition that D^n is a contractive and positively invariant set for the system (3.12), *i.e.*, the condition (3.13) is too conservative for the stability of the system (3.11). We will see in Section 3.5 that condition (3.13) is a special case of one of our results in the present chapter.

2. Systems described by (3.11) have also been used to represent a class of *neural networks* (cf. [59], [87]). It is shown in [87] that neural networks described by (3.11) have certain advantages over the Hopfield model. When considering system (3.11) as a neural network with applications to associative memories, the design objective is to synthesize a system which stores a set of desired patterns as asymptotically stable equilibrium points.

In the application of neural networks to optimization problems (cf. [46], [112]), we wish to construct a network with a unique equilibrium which is globally asymptotically stable, in order to prevent convergence to local minima of an objective function (see also discussions for the continuous-time case in Section 3.1).

3. In many important applications, equation (3.11) may be used to represent *digital processing systems,* including *digital filters* and digital control systems (cf. [6], [7], [8], [11], [12], [24], [25], [31], [33], [36], [51], [61], [61], [67], [89], [90], [91], [92], [98], [102], [103], [106], [107], [108], [114], [120]) with finite wordlength arithmetic under zero external inputs. In such systems, saturation arithmetic is used to cope with the overflow. The absence of limit cycles in such systems is of great interest and can be guaranteed by the global asymptotic stability of the equilibrium $x_e = 0$ for (3.11). The Lyapunov theory has been found to be an appropriate method for solving such problems (cf. [6], [51], [61], [90], [114]). We will review further some of these results in Chapter 6.

In Section 3.5, we establish a general result for the global asymptotic stability of system (3.11). In Section 3.6, we establish results for the global asymptotic stability of system (3.11) by constructing quadratic form Lyapunov functions for the system. In Section 3.7, we consider several specific examples to demonstrate the applicability of the present results. A few pertinent remarks are given in the last section, Section 3.8.

3.5   A General Result for Discrete-Time Systems

In establishing our results, we will make use of Lyapunov functions for the linear systems corresponding to the system (3.11), given by

$$w(k+1) = Aw(k), \quad k = 0, 1, 2, \cdots \qquad (3.14)$$

where $A \in R^{n \times n}$ is defined in (3.11).

In the stability analysis of the equilibrium $x_e = 0$ of system (3.11), we will find it useful to employ Lyapunov functions v whose value for a given state vector $w \notin D^n$ is greater than the value for the corresponding saturated state vector sat(w). Specifically, we will make the following assumption.

Assumption (A– 3.1)   Assume that for the system (3.14), there exists a continuous function $v: R^n \to R$ with the following properties:

(i) v is positive definite, radially unbounded, and

$$Dv_{(3.14)}(w(k)) \triangleq v(w(k+1)) - v(w(k)) = v(Aw(k)) - v(w(k))$$

is negative definite for all $w(k) \in R^n$ (and thus, the eigenvalues of A are within the unit circle); and

(ii) for all $w \in R^n$ such that $w \notin D^n$, it is true that

$$v(\text{sat}(w)) < v(w) \tag{3.15}$$

where D^n and sat($\cdot$) are defined in (3.11). ∎

An example of a function $v_1 : R^2 \to R$ which satisfies (3.15) is given by $v_1(w) = d_1 w_1^2 + d_2 w_2^2$, $d_1, d_2 > 0$. On the other hand, the function $v_2 : R^2 \to R$ given by $v_2(w) = w_1^2 + (2w_1 + w_2)^2$ does not satisfy (3.15). To see this, consider the point $w = [-0.99, 1.05]^T \notin D^2$ and note that $v_2(\text{sat}(w)) = 1.9405$ and $v_2(w) = 1.845$.

We are now in a position to prove the following result.

Theorem 3.2 If Assumption (A–3.1) holds, then the equilibrium $x_e = 0$ of the system (3.11) is globally asymptotically stable.

Since (A–3.1) is true, there exists a positive definite, radially unbounded function v for the system (3.14) such that (3.15) is true, which in turn implies that

$$v(\text{sat}(Aw)) \leq v(Aw), \quad \text{for all } w \in R^n.$$

Also, by (A–3.1),

$$v(Aw(k)) - v(w(k)) < 0, \quad \text{for all } w(k) \neq 0.$$

Thus, along the solutions of the system (3.11), we have

$$Dv_{(3.11)}(x(k)) = v(x(k+1)) - v(x(k)) = v(\text{sat}[Ax(k)]) - v(x(k))$$

$$\leq v(Ax(k)) - v(x(k)) < 0$$

for all $x(k) \neq 0$ and $Dv_{(3.11)}(x(k)) = 0$ if and only if $x(k) = 0$. Therefore, $v(x)$ is positive definite and radially unbounded, and $Dv_{(3.11)}(x)$ is negative definite for all x. Hence, the equilibrium $x_e = 0$ of the system (3.11) is globally asymptotically stable. ∎

Remark 3.4 In particular, for fixed p, $1 \leq p \leq \infty$, let us choose

$$v(w) = \|w\|_p = \left(\sum_{i=1}^{n} |w_i|^p \right)^{1/p}$$

for system (3.14) and assume that $\|A\|_p < 1$, where $\|A\|_p$ denotes the norm induced by $\|w\|_p$. Under these conditions, (A–3.1) is true. To see this, note that v is positive definite and radially unbounded, that

$$v(Aw) = \|Aw\|_p \leq \|A\|_p \|w\|_p < \|w\|_p = v(w),$$

and that

$$\|\mathrm{sat}(w)\|_p < \|w\|_p,$$

for all $w \in R^n$ such that $w \notin D^n$.

Therefore, the equilibrium $x_e = 0$ of the system (3.11) is globally asymptotically stable if

$$\|A\|_p < 1 \qquad (3.16)$$

for some p, $1 \leq p \leq \infty$.

In the case of digital filters, the above argument holds for *any* type of overflow nonlinearity $\varphi: R \to [-1, 1]$. To see this, let

$$f(w) = [\varphi(w_1), \cdots, \varphi(w_n)]^T$$

and note that in this case

$$\|f(w)\|_p < \|w\|_p$$

for all $w \in R^n$ such that $w \notin D^n$. ∎

Remark 3.5 In Section 3.3, we established a stability result for the continuous-time counterpart, given by (3.1), of system (3.11). The condition (3.4) which ensures that the equilibrium $x_e = 0$ of system (3.1) be globally asymptotically stable constitutes a continuous-time counterpart to condition (3.13) (which is a special case of (3.16)). ∎

3.6 Results Involving Quadratic Form Lyapunov Functions

In order to generate quadratic form Lyapunov functions which satisfy Assumption (A–3.1) for systems described by equation (3.11), we will find it convenient to utilize the next assumption.

Assumption (A–3.2) Let

$$x_s = \text{sat}(x) = [\text{sat}(x_1), \cdots, \text{sat}(x_n)]^T$$

for $x \in R^n$ and let $H \in R^{n \times n}$ denote a positive definite matrix. Assume that

$$x_s^T H x_s < x^T H x, \tag{3.17}$$

whenever $x \notin D^n$, $x \in R^n$. ∎

An example of a matrix which satisfies (A–3.2) is any diagonal matrix with positive diagonal elements. On the other hand, the positive definite matrix H given by

$$H = \begin{bmatrix} 5 & 2 \\ 2 & 1 \end{bmatrix},$$

does not satisfy Assumption (A–3.2). (To see this, refer to the example following Assumption (A–3.1) by noting that $v_2(x) = x^T H x$.)

The next result gives a *necessary and sufficient* condition for matrices to satisfy Assumption (A–3.2). This result is very useful in applications.

Lemma 3.1 An $n \times n$ positive definite matrix $H = [h_{ij}]$ satisfies Assumption (A–3.2) *if and only if*

$$h_{ii} \geq \sum_{j=1, j\neq i}^{n} |h_{ij}|, \quad i = 1, \cdots, n. \tag{3.18}$$

This lemma is a special case of Lemma 6.1 (when $L = 1$). See proof on page 63. ∎

The following result is a direct consequence of Theorem 3.2.

Corollary 3.1 The equilibrium $x_e = 0$ of the system (3.11) is globally asymptotically stable, if there exists a matrix H which satisfies (A–3.2), such that

$$Q \triangleq H - A^T H A$$

is positive definite.

By choosing $v(x) = x^T H x$, the proof follows from Theorem 3.2. ∎

Remark 3.6 For *linear* system (3.14), the equilibrium $w = 0$ is globally asymptotically stable if and only if all eigenvalues of A are within the unit circle. Equivalently, the equilibrium $w = 0$ of system (3.14) is globally asymptotically stable if and only if for every positive definite matrix Q, there exists a positive definite matrix P, such that (cf. [93], Theorem 8–17)

$$Q = P - A^T P A. \tag{3.19}$$

Corollary 3.1 tells us that the equilibrium $x_e = 0$ of the *nonlinear* system (3.11) is globally asymptotically stable if in addition to the conditions given above (for linear system (3.14)), Assumption (A–3.2) is satisfied, *i.e.*, there exists a matrix H which satisfies (A–3.2) such that $H - A^T H A$ is positive definite. ∎

In the next result, Theorem 3.3, we show that Corollary 3.1 is actually true when Q is only positive semidefinite, still assuming that A is stable.

Theorem 3.3 The equilibrium $x_e = 0$ of the system (3.11) is globally asymptotically stable, if A is stable and if there exists a matrix H which satisfies (A–3.2), such that

$$Q \triangleq H - A^T H A$$

is positive semidefinite.

Let us choose $v(x(k)) = x^T(k) H x(k)$ for the system (3.11). The function v is clearly positive definite and radially unbounded. Also, since

$$Dv_{(3.11)}(x(k)) = v(x(k+1)) - v(x(k))$$

$$= [\text{sat}(Ax(k))]^T H [\text{sat}(Ax(k))] - x^T(k) H x(k)$$

$$\leq x^T(k)(A^T H A - H)x(k),$$

and since $H - A^T H A$ is positive semidefinite, $Dv_{(3.11)}(x(k))$ is negative semidefinite for all $x(k)$. Therefore, the equilibrium $x_e = 0$ is stable. To show that it is asymptotically stable, we must show that $x(k) \to 0$ as $k \to \infty$.

Let us consider an n consecutive step iteration for the system (3.11), from $n_0 \geq 0$ to $n + n_0$. Without loss of generality, assume that the system (3.11) saturates at $k = l$, $l \in [n_0, n + n_0)$. In view of (A–3.2), it follows that

$$v(x(l+1)) = x^T(l+1) H x(l+1) = [\text{sat}(Ax(l))]^T H [\text{sat}(Ax(l))]$$

$$< [Ax(l)]^T H A x(l) \leq x^T(l) H x(l) = v(x(l)).$$

On the other hand, if no saturation occurs during this period, then, using the fact that if $H - A^T H A$ is positive semidefinite, then $H - (A^T)^n H A^n$ is positive definite when A is stable (cf. [114]), we have

$$v(x(n + n_0)) = x^T(n + n_0) H x(n + n_0) = [A^n x(n_0)]^T H A^n x(n_0)$$

$$= x^T(n_0)(A^T)^n H A^n x(n_0) < x^T(n_0) H x(n_0) = v(x(n_0)).$$

Therefore, we can conclude that for the sequence $\{k: k = 1, 2, \cdots\}$, there always exists an infinite subsequence $\{k_j: j = 1, 2, \cdots\}$, such that $Dv_{(3.11)}(x(k_j))$ is negative for $x(k_j) \neq 0$ and that $v(x(k)) \leq v(x(k_j))$ for all $k \geq k_j$. Since v is a positive definite quadratic form, it follows that $v(x(k_j)) \to 0$ as $j \to \infty$, and therefore $v(x(k)) \to 0$ as $k \to \infty$. This in turn implies that $x(k) \to 0$ as $k \to \infty$. Thus, the equilibrium $x_e = 0$ of (3.11) is globally asymptotically stable. ∎

Remark 3.7 For system (3.11) with A given, the determination of a positive definite matrix H satisfying Assumption (A–3.2) is crucial in the applications of Theorem 3.3 to system (3.11). We will give algorithms for determining a matrix H for a given matrix A in Section 6.5. ∎

3.7 Examples

To demonstrate the applicability of the results in the previous two sections, we now consider two specific examples.

Example 3.1 For the system (3.11) with

$$A = \begin{bmatrix} 1 & 2^{-3} \\ -0.1 & 0.9 \end{bmatrix}, \tag{3.20}$$

we have $\|A\|_p > 1$, $p = 1, 2$, or ∞. Therefore, condition (3.16) fails as global asymptotic stability test for this example.

Hypothesis (A–3.2) is satisfied for this example by choosing

$$H = \begin{bmatrix} 1 & 0.5 \\ 0.5 & 0.8 \end{bmatrix}. \tag{3.21}$$

Since

$$Q = H - A^T H A = \begin{bmatrix} 0.092 & 0.00325 \\ 0.00325 & 0.023875 \end{bmatrix}$$

is positive definite, all conditions of Theorem 3.3 are satisfied and the equilibrium $x_e = 0$ of system (3.11) with A specified by (3.20) is globally asymptotically stable. ∎

Example 3.2 For the system (3.11) with A given by

$$A = \begin{bmatrix} -1 & 0 & 0.1 & 0 \\ 0.2 & -0.6 & 0 & 0.8 \\ -0.1 & 0.1 & 0.8 & 0 \\ 0.1 & 0 & 0.1 & -0.5 \end{bmatrix}, \tag{3.22}$$

it can easily be verified that $\|A\|_p > 1$, $p = 1, 2$, or ∞. Hence, condition (3.16) fails again as a global asymptotic stability test for the present example.

Hypothesis (A–3.2) is satisfied for this example by choosing

$$H = \begin{bmatrix} 1.4 & 0 & -0.2 & 0.4 \\ 0 & 1.6 & 0.2 & -0.4 \\ -0.2 & 0.2 & 3.4 & 0.5 \\ 0.4 & -0.4 & 0.5 & 3 \end{bmatrix}. \tag{3.23}$$

Since

$$Q = H - A^T H A = \begin{bmatrix} 0.026 & 0.161 & -0.003 & 0.077 \\ 0.161 & 1.014 & -0.003 & 0.497 \\ -0.003 & -0.003 & 1.124 & 0.774 \\ 0.077 & 0.497 & 0.774 & 0.906 \end{bmatrix}$$

is positive definite, all conditions of Theorem 3.3 are satisfied, and the equilibrium $x_e = 0$ of the system (3.11) with such a coefficient matrix is globally asymptotically stable. ∎

3.8 Concluding Remarks

In this chapter, we established several new results for the stability analysis of continuous-time and discrete-time dynamical systems with saturation nonlinearities. We believe that these results will be of fundamental interest in practice.

Equation (3.1) describes a class of continuous-time dynamical systems with state saturation nonlinearities—special kinds of hard limiter nonlinearities. Systems of this type arise frequently in the modeling of control systems and neural networks. The stability properties of such systems are of great interest. Our result states that the null solution of system (3.1) will be globally asymptotically stable, if the measure of the coefficient matrix A, induced by the matrix norm $\| \cdot \|_\infty$ is negative. This suggests that matrix measure may play an important role in the stability analysis of systems described by (3.1). Further developments for the stability analysis of continuous-time systems with saturation nonlinearities are expected.

Equation (3.11) describes a large class of discrete-time dynamical systems with saturation nonlinearities (which include important classes of digital filters as special cases). Theorem 3.2 which requires the existence of a function v for system (3.14), satisfying Assumption (A–3.1), guarantees the global asymptotic stability of the equilibrium $x_e = 0$ of system (3.11). The two special forms of the function v considered in this chapter are the l_p vector norm and the quadratic form. These are two important forms of Lyapunov functions. There may be other forms of Lyapunov functions for system (3.14), which satisfy Assumption (A–3.1) under some other conditions.

For the case of quadratic form Lyapunov functions, we established necessary and sufficient conditions under which positive definite matrices can be used to generate these Lyapunov functions for system (3.11) (Lemma 3.1). This renders our result for the global asymptotic stability of the null solution of discrete-time system (3.11) established in Theorem 3.3 very general.

Our results of Sections 3.5 and 3.6 can be directly employed as criteria for testing the non-existence of overflow limit cycles in n^{th} order state-space digital filters using saturation arithmetic. This will be addressed in Chapter 6. We will also see in Chapter 6 that one of our results is in fact the least restrictive criterion for the subject topic.

CHAPTER 4

NULL CONTROLLABILITY OF DISCRETE-TIME DYNAMICAL
SYSTEMS WITH CONTROL CONSTRAINTS AND STATE SATURATION

4.1 Introduction

In the present chapter we will consider discrete-time dynamical systems with symmetric saturation nonlinearities on states and with symmetric constraints on controls. Our current research is distinct from existing research in the following two ways. First, we consider systems with state saturation rather than systems with state constraints. It is generally true that when a state of a system reaches the boundary of a given state constraint set, saturation takes place; and this state saturation may cause loss of the system's stability [4]. Second, we consider a more general problem–controllability, which is also a fundamental requirement in a control system. We note here the difference between a system with state saturation nonlinearities and a system under state constraint: The former is a system with a nonlinear property, while the latter is a system whose states (and/or controls) are not allowed to violate a constraint set.

In this chapter, we investigate the null controllability of discrete-time systems described by

$$x(k + 1) = \text{sat}[Ax(k) + Bu(k)], \quad k = 0, 1, 2, \cdots, \tag{4.1}$$

where $x(k) \in D^n$, $u(k) \in D^m$, $A \in R^{n \times n}$, $B \in R^{n \times m}$, and $\text{sat}(\cdot)$ is defined in (3.11) on page 26. We refer to such systems as "linear systems with control constraints and state saturation". Such systems arise frequently in the modeling of control systems [62] and digital filters [8]. Since we have saturation nonlinearities in (4.1), it is clear that for any $x(0) \notin D^n$, $x(k) \in D^n$, for $k \geq 1$, will always be true. Thus, without loss of generality, we assume that $x(0) \in D^n$. We say that system (4.1) is *null controllable* if for any initial state $x(0) \in D^n$ there exists a *finite* control sequence $u(k) \in D^m, k = 0, 1, 2, \cdots, N$, such that $x(N) = 0$, *i.e.*, the state of system (4.1) can be steered to the origin from any initial state in D^n by constrained controller $u(k) \in D^m$ in a finite number of steps.

Existing results on the null controllability of systems with constrained control are mainly focused on linear systems

$$x(k + 1) = Ax(k) + Bu(k), \tag{4.2}$$

where $A \in R^{n \times n}$, $B \in R^{n \times m}$, $x \in R^n$, $u \in \Omega \subset R^m$ and the control constraint set Ω is convex and compact and contains the origin in its interior. System (4.2) is *globally* null controllable if its state can be steered to the origin by using $u(k) \in \Omega$, in a finite number of steps, for any given initial state $x(0) \in R^n$. It is shown in [55], [109] that, system (4.2) is globally null controllable *if and only if,* (i)

$$\text{rank}[B, AB, \cdots, A^{n-1}B] = n; \tag{4.3}$$

and (ii) every eigenvalue λ of A satisfies $|\lambda(A)| \leq 1$. The results in [55] and [109] have been extended to linear time-varying systems (see [94], [105], and [115]). In the present work we will not consider time-varying systems. Note that here the control constrained set D^m of system (4.1) is also convex and compact and contains the origin in its interior.

We mention here that there exist systems of type (4.1), for which condition (4.3) and $|\lambda(A)| < 1$ do not imply the null controllability in D^n with $u(k) \in D^m$ (see Example 4.4). This suggests a great difference between the qualitative behavior of systems (4.1) and (4.2), since the necessary and sufficient conditions of null controllability for (4.2) do not even constitute sufficient conditions for the null controllability of (4.1). Interestingly, we will also see that it is possible for system (4.1) to be null controllable even when A is not stable (see Example 4.1). It is these observations which motivated the current work.

In Section 4.2, we establish sufficient conditions for the null controllability of system (4.1). We include in Section 4.3 several examples to demonstrate the applicability of the results established in this chapter. In Section 4.4, we conclude the present chapter with a few remarks.

4.2 Main Results

Before presenting our results on the null controllability of system (4.1), let us recall that the equilibrium $x_e = 0$ of a system described by (4.1) under zero-input, given by equation (3.11) (on page 26) is globally asymptotically stable if A is stable and if there exists a matrix H which satisfies Assumption (A–3.2) (on page 31), such that

$$Q \triangleq H - A^T H A$$

is positive semidefinite. This result gives rise to the following definition.

Definition 4.1 A matrix $W \in R^{n \times n}$ is said to possess *Property* $\mathcal{P}$ if it is stable (*i.e.*, all eigenvalues of W are within the unit circle) and if there exists a matrix H satisfying Assumption (A–3.2), such that

$$Q \triangleq H - W^T H W$$

is positive semidefinite. ∎

Thus, the equilibrium $x_e = 0$ of system (3.11) is globally asymptotically stable if A possesses Property $\mathcal{P}$ (see Theorem 3.3 on page 32).

The following two lemmas are required in the proof of our main result (Theorem 4.1).

Lemma 4.1 Suppose $F \in R^{m \times n}$. Then, $Fx \in D^m$ for any $x \in D^n$ *if and only if* $\|F\|_\infty \leq 1$, where $\| \cdot \|_\infty$ denotes the matrix norm induced by the l_∞ vector norm (see definitions in Section 3.2).

Let us denote $F = [f_{ij}]$. The sufficiency is obvious, since

$$\|F\|_\infty \leq 1$$

is equivalent to

$$\sum_{j=1}^n |f_{ij}| \leq 1, \ \ \text{for } i = 1, \cdots, m.$$

Thus for any $x = [x_1, \cdots, x_n]^T \in D^n$, we have

$$-1 \leq -\sum_{j=1}^n |f_{ij}| \leq \sum_{j=1}^n f_{ij} x_j \leq \sum_{j=1}^n |f_{ij}| \leq 1, \ \ \text{for } i = 1, \cdots, m.$$

This is equivalent to

$$Fx \in D^m.$$

To prove the necessity, let us consider the following m vectors in D^n, which are given by

$$X_k = \begin{bmatrix} \text{sign}(f_{k1}) \\ \vdots \\ \text{sign}(f_{kn}) \end{bmatrix}, \quad \text{for } k = 1, \cdots, m.$$

Clearly,

$$FX_k \in D^m \quad \text{for } k = 1, \cdots, m,$$

implies that

$$\sum_{j=1}^{n} |f_{ij}| \leq 1, \quad \text{for } i = 1, \cdots, n.$$

This in turn implies that

$$\|F\|_\infty \leq 1.$$

This proves the lemma. ∎

Lemma 4.2 In system (4.1), if (A, B) is controllable, there exists a set $\Psi \subset (D^n)^0$ with $0 \in (\Psi)^0$, such that every initial point in Ψ can be steered to the origin, with its trajectory remaining inside $(D^n)^0$, in a finite number of steps using $u(k) \in D^m$, where $(\Psi)^0$ denotes the interior of Ψ.

If (A, B) is controllable, the state of system (4.2) can be steered to the origin in n or less steps for any $x(0) \in R^n$ when $u(k)$ is not constrained (cf. [36], [48], and [110]), where n is the order of the system. Let us now constrain the initial state to a set $\Phi \subset D^n$, where Φ is connected and $0 \in (\Phi)^0$. We can find a real number $U > 0$, which depends on Φ and is finite, such that any $x(0) \in \Phi$ can be steered to the origin in n steps with

$$u(k) \in UD^m \triangleq [-U, U]^m.$$

We can assume that $U(\Phi) \geq 1$ for any $\Phi \subset D^n$, without loss of generality. From the continuous dependence of the solutions of (4.2) on the initial state $x(0)$ (cf. [88]), it follows that there exists a real number $r > 1$, such that any initial state

$$x(0) \in \frac{\Phi}{r} \triangleq \left\{ \frac{y}{r} : y \in \Phi \right\}$$

can be steered to the origin with its trajectory within $(D^n)^0$ in n steps by using

$$u(k) \in U_1 D^m,$$

where

$$U_1 = U\left(\frac{\Phi}{r}\right).$$

By choosing

$$\Psi = \frac{1}{U_1} \frac{\Phi}{r} \triangleq \left\{ \frac{x}{rU_1} : x \in \Phi \right\},$$

we can see that $\Psi \subset (D^n)^0$ and that for any $x(0) \in \Psi$, the state of (4.2) can be steered to the origin with its trajectory in $(D^n)^0$ by $u(k) \in D^m$.

The above argument of the existence of Ψ is also valid for system (4.1), since in $(D^n)^0$, system (4.2) and system (4.1) are identical. ∎

We are now in a position to state and prove the next theorem.

Theorem 4.1 System (4.1) is null controllable *if* (A, B) is controllable and *if* there exists an $F \in R^{m \times n}$ with $\|F\|_\infty \leq 1$, such that $G = A + BF$ possesses Property $\mathcal{P}$.

(i) From Lemma 4.1, we know that if $\|F\|_\infty \leq 1$, then $Fx \in D^m$ for any $x \in D^n$. We can substitute the linear state feedback $u(k) = Fx(k)$ into (4.1) to obtain

$$x(k+1) = \text{sat}[Ax(k) + BFx(k)] = \text{sat}[Gx(k)], \quad k = 0, 1, 2, \cdots. \tag{4.4}$$

Since G possesses Property $\mathcal{P}$, we know from Theorem 3.3 that for (4.4), $x_e = 0$ is globally asymptotically stable, *i.e.*, $x(k) \to 0$, as $k \to \infty$ for any initial state $x(0) \in D^n$.

(ii) Since (A, B) is controllable, from Lemma 4.2, we know that there exists a set $\Psi \subset (D^n)^0$ with $0 \in (\Psi)^0$, such that every initial point in Ψ can be steered to the origin, with its trajectory remaining inside $(D^n)^0$, in a finite number of steps using $u(k) \in D^m$.

(iii) Since $x(k) \to 0$, as $k \to \infty$, by using linear state feedback $u(k) = Fx(k)$, the state $x(k)$ can be steered into Ψ in a finite number of steps from any $x(0) \in D^n$. It is clear that on $(D^n)^0$ system (4.1) is equivalent to system (4.2). Therefore the state of system (4.1) can be steered to the origin in a finite number of steps using $u(k) \in D^m$. ∎

Remark 4.1 If the conditions of Theorem 4.1 are satisfied, we have determined a linear state feedback controller which will stabilize the system and for which the resulting feedback control system will never violate the control constraints and will tend to the origin asymptotically. With $u(k) = 0$, it is clear that system (4.2) is stable if A is stable. But for (4.1) with $u(k) = 0$, the system may not be stable even when A is stable, since for some stable A, the state of (4.1) does not go to the origin under zero input for some initial state in D^n (cf. Chapter 3). This indicates the difference between the stabilizability of (4.1) and (4.2). (For results on the stabilization of system (4.2), where the controller is constrained, see, for example, [9], [15], [16], [17], [43], [53], [54], [111], [116].)   ∎

Remark 4.2   The sufficient condition for the null controllability of system (4.1) given in Theorem 4.1 is stated in terms of the feedback law itself. More intrinsic conditions for the null controllability of system (4.1) are not apparent at this time.   ∎

Remark 4.3   In system (4.2), when u is *not constrained*, the system is *controllable* if and only if

$$\text{rank}[\lambda I - A, B] = n \text{ for all complex numbers } \lambda. \tag{4.5}$$

Condition (4.5) is equivalent to condition (4.3). In system (4.2), when u is *not constrained*, the system is *null controllable* if and only if

$$\text{rank}[\lambda I - A, B] = n \text{ for all } nonzero \text{ complex numbers } \lambda \tag{4.6}$$

(see Section 3.3 of [110]).

In this chapter, when we say that (A, B) is controllable, for purposes of simplicity, we mean that A and B satisfy condition (4.3) or (4.5).

We also point out that condition (4.5) (or equivalently, condition (4.3)) implies condition (4.6) while the converse is not true.   ∎

Remark 4.4   The condition that (A, B) be controllable is always required for the null controllability of (4.1); for otherwise we cannot guarantee that the state of the system can be steered to the origin in a finite number of steps using constrained controller $u(k) \in D^m$.
∎

Since system (4.4) is frequently used as the model of a zero-input fixed-point digital filter with saturation arithmetic (cf. Section 6.1), we also have the following results.

Theorem 4.2 System (4.1) is null controllable *if* (A, B) is controllable and *if* one of the following conditions holds:

(i) there exists an $F \in R^{m \times n}$ with $\|F\|_\infty \leq 1$, such that $\rho(|G|) < 1$, where $\rho(\cdot)$ denotes spectral radius, $G = [g_{ij}] = A + BF$, and $|G| = [|g_{ij}|]$; (Recall that $\rho(W) = \max_{1 \leq i \leq n} |\lambda_i(W)|$ for $W \in R^{n \times n}$.)

(ii) there exists an $F \in R^{m \times n}$ with $\|F\|_\infty \leq 1$, such that

$$\|G\|_p = \|A + BF\|_p < 1$$

for some p, $1 \leq p \leq \infty$, where $\|\cdot\|_p$ denotes the matrix norm induced by the l_p vector norm.

(i) Under the condition that $\rho(|G|) < 1$, system (4.4) satisfies $x(k) \to 0$ as $k \to \infty$ for any $x(0) \in D^n$. The proof can be found in [8].

(ii) Choose $v(x) = \|x\|_p$. Then from Remark 3.4 on page 30, we see that the equilibrium $x_e = 0$ of system (4.4) is globally asymptotically stable, *i.e.*, $x(k) \to 0$ as $k \to \infty$ for any $x(0) \in D^n$.

The rest of the proof follows directly from the proof of Theorem 4.1. ∎

Remark 4.5 In $G = A + BF = [g_{ij}]$, we see that

$$g_{ij} = a_{ij} + \sum_{k=1}^{m} b_{ik} f_{kj},$$

$i = 1, 2, \cdots, n$, $j = 1, 2, \cdots, n$. It is clear that $\|A + BF\|_p < 1$ consists of only first order inequalities involving f_{ij} when $p = 1$ or $p = \infty$. Therefore, we can say that $\|A + BF\|_p < 1$ constitutes a "linear" condition when $p = 1$ or $p = \infty$, since in these cases F can be solved by linear programming [117] when $\|F\|_\infty < 1$. We will give a demonstration of this idea in Section 4.3, by applying linear programming to a specific second order system (Example 4.1). ∎

Remark 4.6 Under certain circumstances, the condition

$$\|G\|_p = \|A + BF\|_p < 1$$

in Theorem 4.2 (ii) can be relaxed to

$$\|G\|_p \leq 1.$$

For example, if $p = \infty$, if

$$\max_{i, B_i \neq 0} \left\{ \sum_{j=1}^{n} |g_{ij}| \right\} < 1,$$

if

$$\max_{i, B_i = 0} \left\{ \sum_{j=1}^{n} |g_{ij}| \right\} = \max_{i, B_i = 0} \left\{ \sum_{j=1}^{n} |a_{ij}| \right\} \leq 1,$$

and if (A, B) is controllable, then system (4.1) will still be null controllable. This observation allows us to apply the present results to a larger class of systems (e.g., to systems where (A, B) is in the controllable companion form (see Example 4.3)). ∎

Remark 4.7 The conditions of Theorem 4.1 are less conservative than those in Theorem 4.2 (as can be seen in Example 4.2 of the next section); however, Theorem 4.2 is much easier to apply than Theorem 4.1. ∎

4.3 Examples

We now consider several specific examples to demonstrate the applicability of the present results.

Example 4.1 In system (4.1), take

$$A = \begin{bmatrix} 0.4 & 0.5 \\ -0.3 & -1.2 \end{bmatrix},$$

and

$$B = \begin{bmatrix} 0 \\ 1 \end{bmatrix}.$$

Then

$$A + BF = \begin{bmatrix} 0.4 & 0.5 \\ -0.3 + f_1 & -1.2 + f_2 \end{bmatrix}$$

and $\operatorname{rank}[B \; AB] = 2$.

We can see that there exist many $F \in R^{1 \times 2}$ such that $\|F\|_\infty \leq 1$, i.e., $|f_1| + |f_2| \leq 1$, and $\|A + BF\|_\infty < 1$ (refer to the crosshatched region in Figure 4.1). Therefore, this system is null controllable by Theorem 4.2 (ii).

Note that A is unstable, since $\lambda(A) = -1.1, \; 0.3$. ∎

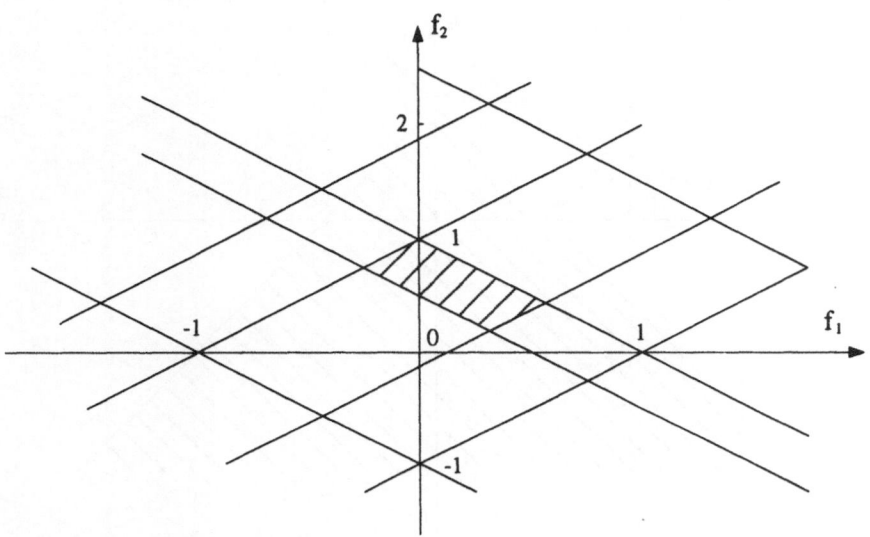

Figure 4.1: The existence of the non-void cross-hatched set in the $f_1 - f_2$ plane ensures the null controllability of the system in Example 4.1

Example 4.2 It is easily seen that for system (4.1) with

$$A = \begin{bmatrix} 0.2 & 1 \\ -1 & 1.8 \end{bmatrix},$$

and

$$B = \begin{bmatrix} 0 \\ 1 \end{bmatrix} \tag{4.7}$$

Theorem 4.2 fails to apply. We attempt to apply Theorem 4.1.

Choosing

$$F = \begin{bmatrix} 0.3 \\ -0.7 \end{bmatrix},$$

we have $\|F\|_\infty = 1$ and

$$G = A + BF = \begin{bmatrix} 0.2 & 1 \\ -0.7 & 1.1 \end{bmatrix}.$$

It can easily be verified that G possesses Property $\mathcal{P}$ for this specific choice of F. To see this, we choose matrix H, which satisfies Assumption (A–3.2), as

$$H = \begin{bmatrix} 1 & -0.6 \\ -0.6 & 1.4 \end{bmatrix}.$$

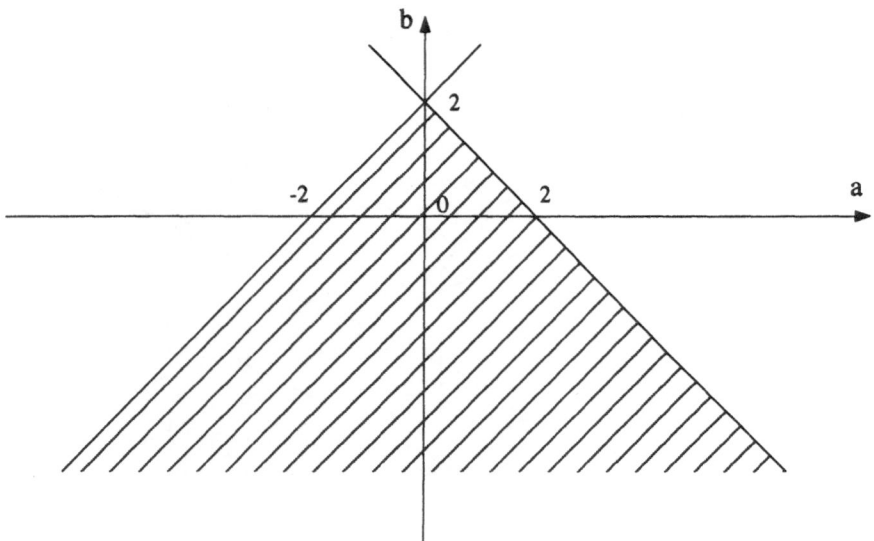

Figure 4.2: Domain of null controllability in the parameter plane obtained by inspection (Example 4.3)

Then,

$$Q = H - A^T H A = \begin{bmatrix} 0.106 & -0.01 \\ -0.01 & 0.026 \end{bmatrix},$$

which is positive definite. Therefore, system (4.1) with such (A, B) is null controllable. ∎

Example 4.3 In the present example, we choose a specific second order case for which the domain of null controllability in the parameter plane can be computed *exactly*, by inspection. We then apply Theorems 4.1 and 4.2 to the same example, to determine the quality of the results obtained by these theorems. *We emphasize that, in general, it is not possible to determine the exact domain of null controllability in the parameter plane for the class of nonlinear systems considered herein.*

We consider the system

$$x(k+1) = \text{sat}[Ax(k) + Bu(k)] = \text{sat}\left[\begin{pmatrix} 0 & 1 \\ b & a \end{pmatrix} x(k) + \begin{pmatrix} 0 \\ 1 \end{pmatrix} u(k) \right]. \tag{4.8}$$

a) When $u(k)$ is constrained to $[-1, 1]$, it can be shown by inspection that system (4.8) is null controllable when (a, b) belongs to the crosshatched region shown in Figure 4.2.

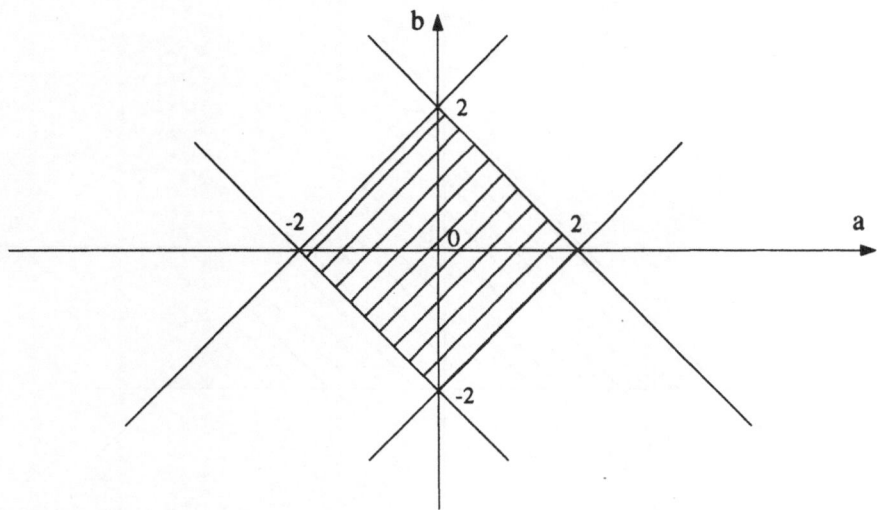

Figure 4.3: Domain of null controllability in the parameter plane obtained by Theorem 4.2 (ii) when $p = \infty$ (Example 4.3)

b) For conditions

$$\|A + BF\|_{p=\infty} < 1$$

and

$$\|F\|_\infty \leq 1$$

to be satisfied, (a, b) must belong to the crosshatched region indicated in Figure 4.3, which is the region of null controllability for system (4.8), obtained by Theorem 4.2 (ii), when $p = \infty$ (cf. Remark 4.6).

c) For the conditions of Theorem 4.1 to be satisfied, (a, b) must belong to the crosshatched region shown in Figure 4.4 (cf. Section 6.3 for details).

Note that the crosshatched region in Figure 4.4 contains the crosshatched region of Figure 4.3 and is contained in the crosshatched region of Figure 4.2. ∎

Example 4.4 In the present example, we demonstrate the validity of our original claim that for system (4.1) the conditions that A be stable and (A, B) be controllable do not imply

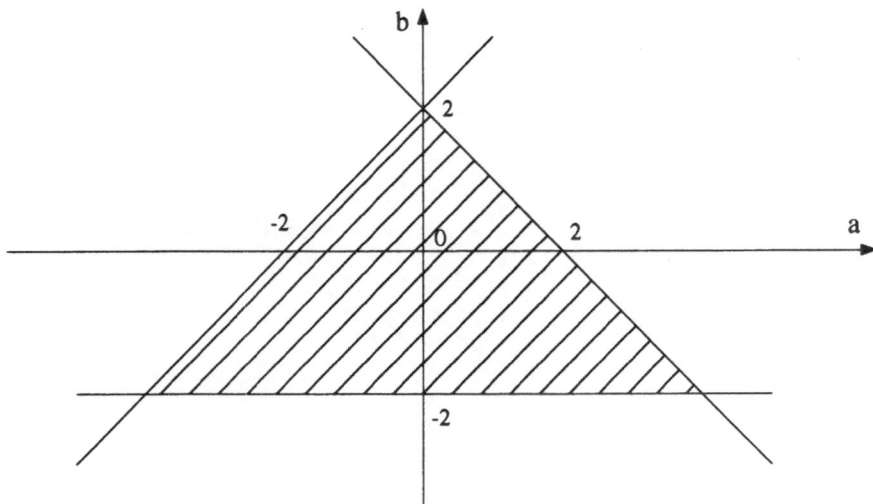

Figure 4.4: Domain of null controllability in the parameter plane obtained by Theorem 4.1 (Example 4.3)

null controllability. Specifically, for the system

$$x(k+1) = \text{sat}[Ax(k) + Bu(k)] = \text{sat}\left[\begin{pmatrix} 1.6 & 0.2 \\ -30 & -3.3 \end{pmatrix} x(k) + \begin{pmatrix} 0 \\ 1 \end{pmatrix} u(k)\right],$$

where $x(0) \in D^2 = [-1, 1]^2$ and $u(k) \in D^1 = [-1, 1]$, we have

$$\text{rank}[B \ AB] = 2$$

and $\lambda_1(A) = -0.8$ and $\lambda_2(A) = -0.9$. However, this system is not null controllable, since when we choose

$$x(0) = \begin{bmatrix} a \\ b \end{bmatrix} \quad \text{or} \quad \begin{bmatrix} -a \\ b \end{bmatrix},$$

with $0.75 \leq a \leq 1$ and $-1 \leq b \leq 1$, the state of this system will remain at

$$\begin{bmatrix} 1 \\ -1 \end{bmatrix} \quad \text{or} \quad \begin{bmatrix} -1 \\ 1 \end{bmatrix}$$

respectively, for any $u \in D^1$. ∎

PART II
STABILITY ANALYSIS OF ONE-DIMENSIONAL AND
MULTIDIMENSIONAL STATE-SPACE DIGITAL FILTERS WITH
OVERFLOW NONLINEARITIES

CHAPTER 5

INTRODUCTION TO PART II

5.1 Fixed-Point Digital Filters and Overflow Nonlinearities

In most realizations of digital filters, signals are encoded in a particular format (mostly binary format in which each number is represented by a sign bit and a magnitude) and are stored in registers which have a finite wordlength. Irrespective of the encoding and registering of the signals, multiplications and additions generally lead to an increase in the wordlength required for the results of the operations. This increase of requirement on the wordlength usually results in an overflow which requires special algorithms to cope with, since a wordlength reduction is necessary to prevent the wordlength of signals from increasing indefinitely. These are typical situations in a digital filter realized using finite wordlength format, or *fixed-point* algorithm. Another effect existing in fixed-point digital filters is quantization, which is a consequence of a finite number of representations in fixed-point algorithms. In our research, we do not consider quantization effects and consider only overflow nonlinearities in the fixed-point state-space digital filters. Although one cannot completely eliminate quantization errors when studying overflow effects, it has been noted that quantization effects can be made arbitrarily small by decreasing the quantization step size [26].

As mentioned above, a fixed-point algorithm results in a finite range number representation, and multiplications and additions pose a problem when a result of these operations falls outside the representable range. An *overflow* nonlinearity results when this number is modified so that it falls back within the representable range. In general, the overflow nonlinearity changes the most significant bits as well as the least significant bits of a fixed-point number. These types of nonlinearities are well described in the literature (see, e.g., Erickson

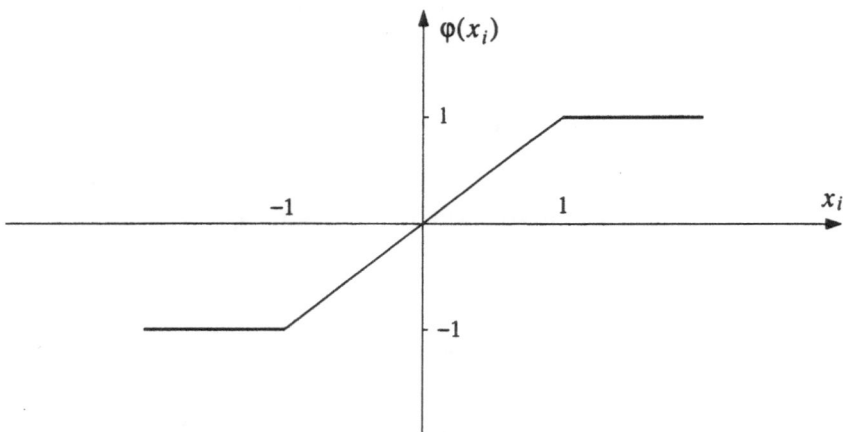

Figure 5.1: Saturation nonlinearity

and Michel [33]) and, therefore, will only be briefly discussed here. Several types of arithmetic have been used extensively in practice to cope with the overflow in fixed-point digital filters. If an overflow occurs, a number of different actions may be taken. If the number that causes the overflow is replaced by a number having the same sign, but with a magnitude corresponding to the overflow level, a *saturation overflow* characteristic shown in Figure 5.1 is obtained. *Zeroing overflow* substitutes the number zero in case of an overflow (see Figure 5.2). In two's complement arithmetic, the most significant bits that cause the overflow are discarded. In this case, overflows in intermediate results do not cause errors, as long as the final result does not have overflow. This *two's complement overflow* characteristic is illustrated in Figure 5.3. Another way of dealing with overflow is the *triangular overflow* characteristic (see Figure 5.4). We note that there are other types of overflow arithmetic in use in current technology.

It is possible to have different wordlengths for various signals in a filter, resulting in different overflow levels. We will assume throughout this chapter and the next two chapters that all overflow nonlinearities in a given filter have the same overflow level (normalized to 1) and are of the same type.

In the literature [33], overflow nonlinearities are usually viewed as belonging to a sector

4.4 Concluding Remarks

Because of wide applications of system (4.1) in *control systems* and *fixed-point digital filters*, the null controllability of such systems is of great interest. We emphasize that when the controller is constrained (as in the present case, where $u(k) \in D^m$), system (4.1) may not be null controllable even when (A, B) is controllable and A is stable.

We solve the problem addressed herein in two steps. (i) We give conditions under which system (4.1) can be stabilized by constrained controllers $u(k) \in D^m$; and (ii) We prove that if system (4.1) is stabilizable by constrained controllers and if (A, B) is controllable, then system (4.1) is null controllable. As pointed out (Remark 4.1), our results (Theorems 4.1 and 4.2) can also serve as conditions for *stablizability* of system (4.1), in which case we do not require that (A, B) be controllable.

The present results are new. Further investigations are expected to establish more general results for the null controllability of system (4.1), which will be independent of the stabilizability of system (4.1).

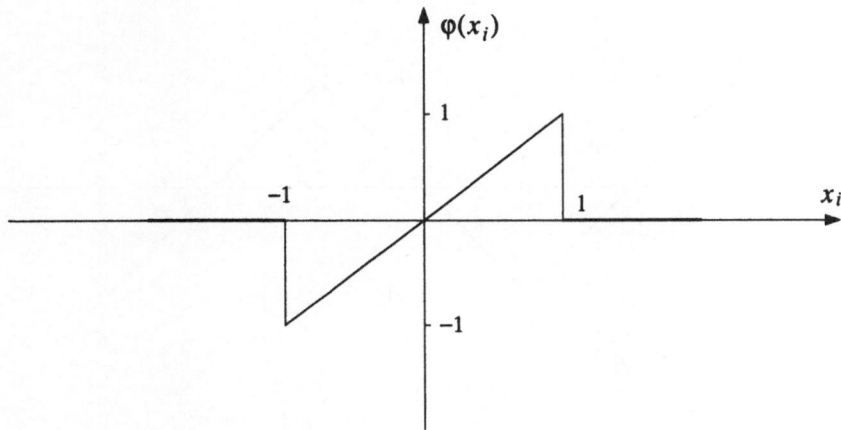

Figure 5.2: Zeroing arithmetic

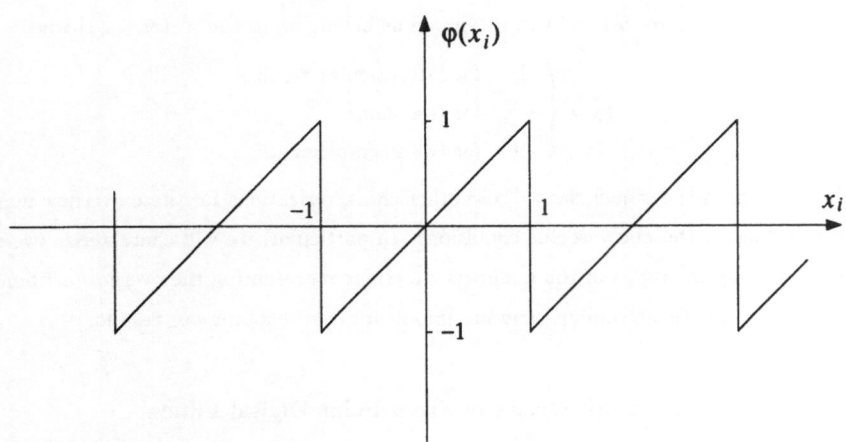

Figure 5.3: Two's complement

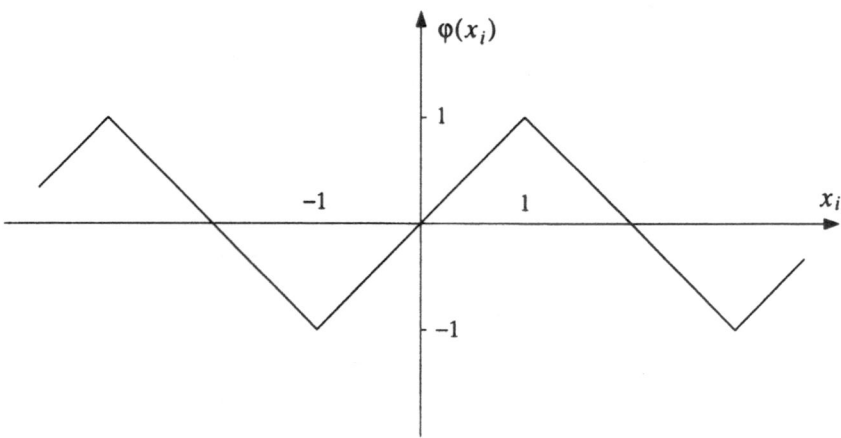

Figure 5.4: Triangular characteristic

$[k_m, k_M]$. Thus if $\varphi(\cdot)$ denotes a given nonlinearity, then

$$k_m \sigma^2 \leq \sigma\varphi(\sigma) \leq k_M \sigma^2, \quad \text{for all} \;\; \sigma \in R$$

where k_m, k_M are constants such that $-\infty < k_m \leq k_M < \infty$. For example, the overflow nonlinearities in Figures 5.1–5.4 can be viewed as belonging to the sector $[k_0, 1]$ where

$$k_0 = \begin{cases} 0, & \text{for saturation or zeroing} \\ -\frac{1}{3}, & \text{for triangular} \\ -1, & \text{for two's complement.} \end{cases}$$

As will be seen in the sequel, we will use other characterizations for these overflow nonlinearities instead of the above sector conditions. In particular, we will characterize overflow nonlinearities by the range of the nonlinear functions representing the overflow arithmetic. Our characterizations are unique, and are important in developing our results.

5.2 Limit Cycles in Fixed-Point Digital Filters

Effects of overflow in digital filters, such as limit cycles (overflow oscillations), have been known for a long time. Any nontrivial periodic solution of a fixed-point digital filter is called a limit cycle. It is customary in the study of digital filters also to view nonzero equilibrium points as limit cycles. Unless otherwise stated, we will follow this practice. It is well-known that overflow corrections in digital filters can result in large amplitude limit cycles, even though the filter is linearly stable. Determination for the non-existence of limit cycles in fixed-point digital filters is still an open problem. Various criteria have been developed thus far, with many of them using the idea that the global asymptotic stability of a filter's null solution guarantees the non-existence of limit cycles.

The study of limit cycles in fixed-point digital filters began in the late 60's. The first important contribution is probably that given by Sandberg [102] and by Ebert *et al* [31] in 1969. For the zero-input response of a *second-order direct form* linearly stable digital filter employing the saturation arithmetic, it is shown in [102] that the amplitude of all limit cycles can be made arbitrarily small by increasing the number of bits used in the representation of the data samples when roundoff type quantization is taken into account, and in [31] it is shown that limit cycles cannot exist when quantization is neglected. The works in [102] and [31] are important because the cascade and parallel realizations of any order direct form digital filters are both formed using second-order sections as building blocks.

Direct form digital filters are restricted to single-input and single-output cases. For filters with multiple inputs and/or multiple outputs, one has to use state-space representations. For n^{th} order state-space digital filters endowed with overflow nonlinearities, the first significant contribution was made by Barnes and Fam [6]. In [6], it is shown that any (state-space) digital filter that is linearly stable and uses an arithmetic for which overflow does not increase the Euclidean norm of the state vector, can be realized in a form that will be free of limit cycles. The idea of finding conditions under which the overflow nonlinearity does not increase the Euclidean norm of the state vector is very important. It has evoked several other contributions in the literature, such as [90] and [114]. In [90] and [114], a generalized form of Euclidean norm of the state vector, which is also called the *quadratic norm*, is employed, and further relaxations of the result in [6] are also achieved (for details, see Section 6.2 of this monograph).

An effective method of analyzing the stability of fixed-point digital filters (when we do not consider quantization effects) is Lyapunov's Second Method introduced in Chapter 2. Results established using the Lyapunov method (for the one-dimensional case) are usually called *time domain criteria* (e.g., results in [6], [90] and [114]). There are also some existing *frequency domain criteria* which are generally complicated in form and difficult to apply. One of the goals in this part of our research (Chapter 6) is to establish new (time domain) criteria for the non-existence of limit cycles in n^{th} order state-space fixed-point digital filters. Our results will be derived using Lyapunov's Second Method.

Our work in Chapter 6 can be viewed as further developments of the results of Chapter 3 (for the discrete-time case) and it can also be viewed as a further development of the work in [6], [90] and [114]. Results established herein constitute significant improvements over the results in [6], [90] and [114].

5.3 Multidimensional Digital Filters

In applications of digital signal processing, there are many occasions to deal with signals which are not in the time domain and are not one-dimensional (such as digital image processing). Digital image processing is a typical area of multidimensional signal processing in which the signals are represented in a two-dimensional spatial domain. Multidimensional signal processing and one-dimensional signal processing differ in a variety of ways. For example, multidimensional problems generally involve considerably more data than one-dimensional ones and the mathematics for handling multidimensional systems is less completely developed than the mathematics for handling one-dimensional systems. As in the one-dimensional case, multidimensional filters are also implemented using fixed-point algorithms (finite wordlength format), and of course, we have to deal with overflow phenomena in these filters. The usual types of arithmetic used to cope with overflow in multidimensional filters are identical to those used in one-dimensional filters (*i.e.*, the saturation, zeroing, two's complement, triangular, etc.).

Several ways are commonly used to represent the operations in multidimensional signal processing. These include transfer functions and *partial difference equations*. We are particularly interested in a class of multidimensional digital filters described by state-space (partial) difference equations, known as Roesser's model [99] with overflow nonlinearities. The model considered herein is also called the *local state-space realization* and is regarded as the multidimensional counterpart to the one-dimensional state-space realization. It is a basic requirement to have criteria for choosing the parameters of a filter so that the null solution of the filter is globally asymptotically stable, which implies the non-existence of limit cycles in the filter. This has been a topic of increasing interest during the past few years (cf. [1], [2], [7], [8], [64], [69], [113]).

The study of stability properties of multidimensional digital filters endowed with overflow nonlinearities started in the late 70's. The first criterion which sufficiently guarantees the global asymptotic stability of state-space two-dimensional digital filters with overflow nonlinearities was given by El-Agizi and Fahmy [32] in 1979. The results in [32] are generalizations of the one-dimensional results in [6] and [90] to the two-dimensional case. A few other results related to this topic were established in [2], [7], [8], and [113], which will be further reviewed in Chapter 7.

It is believed that Lyapunov's Second Method can become a powerful tool in the stability analysis of multidimensional digital filters. The Lyapunov method was originally developed for systems which are described in the time domain. As will be seen in Chapter 7, after some adaptations, the Lyapunov method can also be applied to the stability analysis of spatial domain systems, as in our case, to multi-dimensional state-space digital filters with overflow nonlinearities. One of our results in Chapter 7 constitutes a relaxation to a result in [32]. The results in Chapter 7 are also generalizations of our results in Chapter 6, from the one-dimensional case, to the multidimensional case.

CHAPTER 6

CRITERIA FOR THE ABSENCE OF OVERFLOW OSCILLATIONS IN FIXED-POINT DIGITAL FILTERS USING GENERALIZED OVERFLOW CHARACTERISTICS

6.1 Introduction

Equations of the form (cf. (3.11))

$$x(k+1) = \text{sat}[Ax(k)], \quad k = 0, 1, 2, \cdots \tag{6.1}$$

may also be employed to represent fixed-point digital filters using saturation overflow arithmetic under zero input. This model does not include quantization effects.

The existence and non-existence of limit cycles (nonlinear oscillations) in digital filters (under zero input) due to overflow nonlinearities have been investigated extensively during the last two decades (see, e.g., [6], [7], [8], [11], [12], [24], [25], [31], [33], [51], [61], [67], [90], [91], [92], [98], [102], [103], [106], [107], [108], [114], [120]). The types of characteristics considered in these studies include zeroing, two's complement, triangular, saturation, and other types of overflow nonlinearities. Since stable *second-order* direct form digital filters using saturation arithmetic have been shown to be free of limit cycles (cf. [31], [102]), filters of *any order*, endowed with saturation nonlinearities have received special attention. In addition, as pointed out in [92], [103], [108], conditions obtained for the absence of nonlinear oscillations (under zero input) in digital filters with saturation overflow nonlinearities are generally less conservative than corresponding conditions obtained for digital filters with other types of overflow characteristics.

In Section 6.2 of the present chapter, we review some of the existing criteria for the absence of limit cycles in fixed-point digital filters. We utilize the results of Chapter 3 for the asymptotic stability of discrete-time systems with saturation nonlinearities to establish in Section 6.3 new conditions for the non-existence of limit cycles for digital filters (6.1). In Section 6.4, we consider digital filters with generalized overflow nonlinearities and establish conditions for the non-existence of limit cycles in such filters. We develop algorithms which can be used to determine the required positive definite matrices for generating quadratic form Lyapunov functions in Section 6.5. We give several specific examples to demonstrate the applicability of the present results in Section 6.6 and we make a few pertinent remarks in Section 6.7.

6.2 Some Existing Results

An early result of Barnes and Fam [6] states that if

$$\|A\|_2 = \sqrt{\lambda_M(A^TA)} < 1 \tag{6.2}$$

where $\|\cdot\|_2$ denotes the norm of a matrix induced by the l_2 vector norm, A^T represents the transpose of A, and $\lambda_M(A^TA)$ denotes the maximum eigenvalue of A^TA, then the digital filter (6.1) is free of limit cycles. It turns out that this result is true for many other types of overflow nonlinearities, including zeroing, two's complement, triangular, and saturation characteristics.

An extension to this result is that the matrix Q in (6.3) be positive semidefinite, assuming that A is stable, *i.e.*, assuming that every eigenvalue λ_i of A satisfies $|\lambda_i| < 1$,

$$Q = D - A^TDA \geq 0, \tag{6.3}$$

where D is a diagonal matrix with positive diagonal elements [51], [90], [114]. This result can not be applied to the case in which the absolute values of some diagonal elements in matrix A are greater than or equal to 1.

Another extension to condition (6.2) is given by

$$\|A\|_p < 1, \quad \text{for some} \quad p, \ \ 1 \leq p \leq \infty \tag{6.4}$$

where $\|\cdot\|_p$ denotes the matrix norm induced by the l_p vector norm. (Condition (6.4) is stated in [12] without proof.)

Another time domain result states that if

$$\rho(|A|) < 1, \tag{6.5}$$

62

where $\rho(\cdot)$ denotes the spectral radius (see definition on page 42) and $|A| = [|a_{ij}|]$, then the digital filter (6.1) is free of limit cycles (cf. [7], [8], [12]). This result is especially useful for testing a digital filter with lower or upper triangular coefficient matrix. It was shown recently [50] that condition (6.5) is a special case of condition (6.3).

It is shown by Singh [107], [108], that the condition

$$D + DA(zI - A)^{-1} + [DA(zI - A)^{-1}]^* \geq 0, \quad \text{for all } |z| = 1, \tag{6.6}$$

is a frequency domain equivalent condition to (6.3), where I denotes the $n \times n$ identity matrix, z is a complex variable, and $*$ represents the conjugate transpose.

An improvement to condition (6.6), assuming saturation arithmetic in the digital filters, given by

$$2D + DA(zI - A)^{-1} + [DA(zI - A)^{-1}]^* \geq 0, \quad \text{for all } |z| = 1, \tag{6.7}$$

is also due to Singh [108]. Note that in (6.6) and (6.7), D is still assumed to be a diagonal matrix with positive diagonal elements and that A is assumed to be a stable matrix.

We note that conditions (6.2)–(6.7) constitute also conditions for the global asymptotic stability of the null solutions of the digital filters under investigation (with no external inputs).

6.3   Digital Filters Using Saturation Arithmetic

Since no limit cycles can exist in a digital filter if its trivial solution is globally asymptotically stable, we can use the results of Sections 3.5 and 3.6, to establish the following results for n^{th} order digital filters with saturation arithmetic.

Corollary 6.1

(i) A digital filter described by (6.1) (or by (3.11)) is free of limit cycles, if Assumption (A–3.1) (on page 28) is satisfied.

(ii) A digital filter described by (6.1) (or by (3.11)) is free of limit cycles, if A is stable and if there exists a matrix H which satisfies Assumption (A–3.2) (on page 31), such that

$$Q \triangleq H - A^THA$$

is positive semidefinite.

Remark 6.1 As pointed out in Remark 3.4, condition (6.4) is a special case of Corollary 6.1 (i). We also note that since in (6.3), D is assumed to be a diagonal matrix with positive diagonal elements, Corollary 6.1 (ii) constitutes a generalization of condition (6.3). ∎

Remark 6.2 The results given in Corollary 6.1 are in general less conservative than conditions (6.2)–(6.6) and appear to have no direct relationships with condition (6.7). However, Corollary 6.1 (ii) is considerably easier to apply than condition (6.7), since the latter involves the inversion of a matrix of variables. In Section 6.6, we include a specific example which can be analyzed by Corollary 6.1 (ii), but not by any of the previous results given by conditions (6.2)–(6.7). ∎

Remark 6.3 In [98], it is shown that second-order digital filters given by (6.1) with

$$A = \begin{bmatrix} a_{11} & a_{12} \\ a_{21} & a_{22} \end{bmatrix}, \tag{6.8}$$

are free of limit cycles if A is stable and if

$$|a_{11} - a_{22}| \leq 2\min(|a_{12}|, |a_{21}|) + 1 - \det(A). \tag{6.9}$$

This result can also be derived by Corollary 6.1 (ii), since under the above conditions, there always exists a matrix H which satisfies Assumption (A–3.2) with $H - A^T H A$ positive semidefinite (cf. [98]).

We also note that when for a second-order digital filter with

$$A = \begin{bmatrix} 0 & 1 \\ b & a \end{bmatrix},$$

the parameters (a, b) are located within the well-known stability triangle, then condition (6.9) is automatically satisfied. Thus, second-order direct form digital filters with saturation nonlinearities and with matrix A stable, are free of limit cycles. This result was originally established in [31] and [102], using approaches which differ significantly from the present method. ∎

6.4 Digital Filters Using Generalized Overflow Characteristics

In the remainder of this chapter, we will consider n^{th} order digital filters described by equations of the form

$$x(k+1) = f[Ax(k)], \quad k = 0, 1, 2, \cdots \tag{6.10}$$

where $x(k) \in R^n$, $A \in R^{n \times n}$,

$$f(x) = [\varphi(x_1), \varphi(x_2), \cdots, \varphi(x_n)]^T, \tag{6.11}$$

and $\varphi: R \to [-1, 1]$ is piecewise continuous. We call system (6.10) a *fixed-point digital filter using overflow arithmetic*. For such filters, we will make the following assumption.

Assumption (A–6.1)   Let f be defined as in (6.11). Assume that $H \in R^{n \times n}$ is a positive definite matrix and that

$$f(x)^T H f(x) < x^T H x, \tag{6.12}$$

for all $x \in R^n$, $x \notin D^n$. ∎

In what follows, we will let the function φ in (6.11) be defined as (see Figure 6.1)

$$\varphi(x_i) = \begin{cases} L, & x_i > 1 \\ x_i, & -1 \le x_i \le 1 \\ -L, & x_i < -1 \end{cases} \tag{6.13}$$

or (see Figure 6.2)

$$\begin{cases} L \le \varphi(x_i) \le 1, & x_i > 1 \\ \varphi(x_i) = x_i, & -1 \le x_i \le 1 \\ -1 \le \varphi(x_i) \le -L, & x_i < -1 \end{cases} \tag{6.14}$$

where $-1 \le L \le 1$. We will call the function φ defined in (6.13) and (6.14) a *generalized overflow characteristic*. Note that when defined in this way, the function φ includes as special cases the usual types of overflow arithmetic employed in practice, such as zeroing, two's complement, triangular, and saturation overflow characteristics.

To establish our next result, Theorem 6.1, we require the following preliminary result, Lemma 6.1.

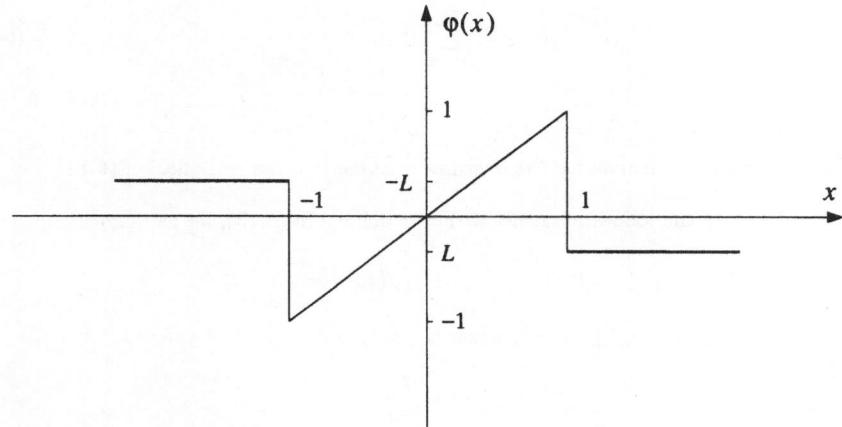

Figure 6.1: The generalized overflow nonlinearity described by (6.13)

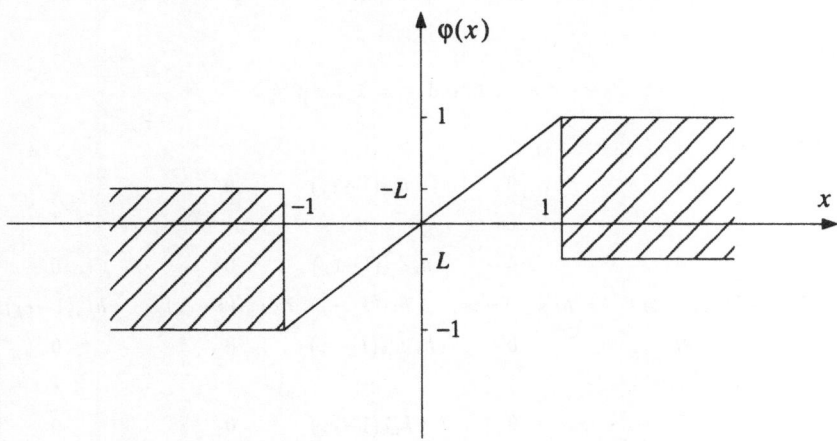

Figure 6.2: The generalized overflow nonlinearity described by (6.14)

Lemma 6.1 Assume that f is defined in (6.11) and φ is given in (6.13) or in (6.14) with $-1 < L \leq 1$. An $n \times n$ positive definite matrix $H = H^T = [h_{ij}]$ satisfies Assumption (A-6.1) *if and only if*

$$(1+L)h_{ii} \geq 2 \sum_{j=1, j \neq i}^{n} |h_{ij}|, \quad i = 1, \cdots, n. \tag{6.15}$$

We first prove this lemma for the overflow arithmetic given in Equation (6.13).

We introduce the following notation. For φ defined in (6.13), let us denote

$$f(x) = [\varphi(x_1), \cdots, \varphi(x_n)]^T = Ex$$

where $E = \text{diag}[e_1, e_2, \cdots, e_n]$, $e_i = 1$ when $|x_i| \leq 1$, and

$$e_i = \frac{L}{|x_i|}$$

when $|x_i| > 1$. Then, we have

$$x^T H x - f(x)^T H f(x) = x^T (H - EHE) x.$$

Sufficiency: Suppose

$$x = [x_1, x_2, \cdots, x_n]^T, \ |x_k| > 1 \ \text{and} \ |x_i| \leq 1 \ \text{for} \ i \neq k \ (x \notin D^n).$$

We have

$$-1 < e_k < 1 \ \text{and} \ e_i = 1 \ \text{for} \ i \neq k,$$

and therefore,

$$H - EHE = \begin{bmatrix} 0 & \cdots & 0 & h_{1k}(1-e_k) & 0 & \cdots & 0 \\ \vdots & \cdots & \vdots & \vdots & \vdots & \cdots & \vdots \\ 0 & \cdots & 0 & h_{k-1,k}(1-e_k) & 0 & \cdots & 0 \\ h_{k1}(1-e_k) & \cdots & h_{k,k-1}(1-e_k) & h_{kk}(1-e_k^2) & h_{k,k+1}(1-e_k) & \cdots & h_{kn}(1-e_k) \\ 0 & \cdots & 0 & h_{k+1,k}(1-e_k) & 0 & \cdots & 0 \\ \vdots & \cdots & \vdots & \vdots & \vdots & \cdots & \vdots \\ 0 & \cdots & 0 & h_{nk}(1-e_k) & 0 & \cdots & 0 \end{bmatrix}$$

and

$$x^T(H - EHE)x = (1 - e_k)\Big(h_{kk}(1 + e_k)x_k^2 + 2 \sum_{i=1, i \neq k}^{n} h_{ik}x_i x_k\Big). \tag{6.16}$$

Note that in the above equation we have used the fact that $h_{ij} = h_{ji}$.

From $|x_i| \leq 1$ for $i \neq k$, $|x_k| > 1$, $e_k|x_k| = L$ and $L > -1$, we have

$$(1 + L)|x_i x_k| \leq (1 + L)|x_k| < (|x_k| + L)|x_k| = (1 + e_k)x_k^2.$$

Hence, from (6.16), we have

$$x^T(H - EHE)x \geq (1 - e_k)\left(h_{kk}(1 + e_k)x_k^2 - 2\sum_{i=1, i \neq k}^{n} |h_{ik}x_i x_k|\right)$$

$$> (1 - e_k^2)x_k^2\left(h_{kk} - \frac{2}{1 + L}\sum_{i=1, i \neq k}^{n} |h_{ik}|\right) \geq 0,$$

i.e.,

$$x^T H x > x^T E H E x = f(x)^T H f(x).$$

Denote

$$M = \{1, 2, \cdots, m\}$$

for *any m*, $0 < m \leq n$, and

$$N = \{k_i : 0 < k_i \leq n, \; k_i \neq k_j, \; \text{when } i \neq j, \; i \in M\}.$$

Now suppose that

$$x = [x_1, x_2, \cdots, x_n]^T, \; |x_k| > 1 \; \text{ for } \; k \in N$$

and

$$|x_i| \leq 1 \; \text{ for } \; i \notin N \; (x \notin D^n).$$

Following the same procedure as above, we have

$$x^T(H - EHE)x = \sum_{k \in N}(1 - e_k)\left(h_{kk}(1 + e_k)x_k^2 + 2\sum_{i=1, i \notin N}^{n} h_{ik}x_i x_k\right)$$

$$+ \sum_{k \in N}\sum_{l \in N, l \neq k} h_{kl}x_k x_l(1 - e_k e_l)$$

$$\geq \sum_{k \in N}(1 - e_k)\left(h_{kk}(1 + e_k)x_k^2 - 2\sum_{i=1, i \notin N}^{n} |h_{ik}x_i x_k|\right)$$

$$+ \sum_{k \in N}\sum_{l \in N, l \neq k} h_{kl}x_k x_l(1 - e_k e_l)$$

$$> \sum_{k \in N}(1 - e_k^2)x_k^2\left(h_{kk} - \frac{2}{1 + L}\sum_{i=1, i \notin N}^{n} |h_{ik}|\right)$$

$$+ \sum_{k \in N} \sum_{l \in N, l \neq k} h_{kl} x_k x_l (1 - e_k e_l)$$

$$= \sum_{k \in N} (1 - e_k^2) x_k^2 \left(h_{kk} - \frac{2}{1 + L} \sum_{i=1, i \neq k}^{n} |h_{ik}| \right)$$

$$+ \frac{2}{1 + L} \sum_{k \in N} (1 - e_k^2) x_k^2 \sum_{i \in N, i \neq k} |h_{ik}|$$

$$+ \sum_{k \in N} \sum_{l \in N, l \neq k} h_{kl} x_k x_l (1 - e_k e_l). \tag{6.17}$$

The first summation of the right hand side in (6.17) is nonnegative, by assumption. Considering the last two terms in (6.17), by noting that $-1 < e_k < 1$ and $e_k |x_k| = L$ for $k \in N$, and $-1 < L \leq 1$, we have

$$\frac{2}{1 + L} \sum_{k \in N} (1 - e_k^2) x_k^2 \sum_{i \in N, i \neq k} |h_{ik}| + \sum_{k \in N} \sum_{l \in N, l \neq k} h_{kl} x_k x_l (1 - e_k e_l)$$

$$\geq \sum_{k \in N} \sum_{l \in N, l \neq k} (1 - e_k^2) x_k^2 |h_{kl}| - \sum_{k \in N} \sum_{l \in N, l \neq k} |h_{kl} x_k x_l| (1 - e_k e_l)$$

$$= \sum_{k \in N} \sum_{l \in N, l \neq k} |h_{kl} x_k| (|x_k| - e_k L - |x_l| + e_k L)$$

$$= \sum_{k \in N} \sum_{l \in N, l \neq k} |h_{kl}| x_k^2 - \sum_{k \in N} \sum_{l \in N, l \neq k} |h_{kl} x_k x_l|$$

$$= \sum_{k \in N} \sum_{l \in N, l > k} |h_{kl}| (x_k^2 + x_l^2) - 2 \sum_{k \in N} \sum_{l \in N, l > k} |h_{kl} x_k x_l|$$

$$= \sum_{k \in N} \sum_{l \in N, l > k} |h_{kl}| (|x_k| - |x_l|)^2 \geq 0.$$

Therefore,

$$x^T H x - f(x)^T H f(x) = x^T (H - EHE) x > 0,$$

for any $x \in R^n$ such that $x \notin D^n$.

This proves the sufficiency.

Necessity: It suffices to show that if (6.15) does not hold, there always exist some points $x \notin D^n$, such that

$$x^T H x \leq f(x)^T H f(x).$$

Suppose that (6.15) does not hold for $i = k$, i.e.,

$$\delta \stackrel{\triangle}{=} 2 \sum_{j=1, j \neq k}^{n} |h_{kj}| - (1 + L) h_{kk} > 0.$$

Let us choose

$$|x_k| = 1 + \xi, \ \xi > 0, \ \text{and} \ x_i = -sign(h_{ik}x_k), \ i \neq k,$$

where

$$sign(y) = \begin{cases} 1, & y > 0 \\ 0, & y = 0 \\ -1, & y < 0 \end{cases}.$$

Then, $x = [x_1, \cdots, x_n]^T \notin D^n$ and (6.16) becomes

$$x^T(H - EHE)x = (1 - e_k)\left(h_{kk}(1 + e_k)x_k^2 - 2 \sum_{i=1,i\neq k}^{n} |h_{ik}x_k|\right)$$

$$= (1 - e_k)|x_k|\left(h_{kk}\xi + (1 + L)h_{kk} - 2 \sum_{i=1,i\neq k}^{n} |h_{ki}|\right)$$

$$= (1 - e_k)|x_k|(h_{kk}\xi - \delta).$$

Clearly, when we choose

$$0 < \xi \leq \frac{\delta}{h_{kk}},$$

we have

$$x^T H x - f(x)^T H f(x) = x^T(H - EHE)x \leq 0.$$

Note here that $h_{kk} > 0$ since H is positive definite.

This proves the necessity.

For the overflow nonlinearity given in (6.14), the proof of sufficiency is similar to the proof given above. To prove necessity, we note that for a given L, when $|x_i| > 1$, $\varphi(x_i)$ in (6.14) may assume *any* value in the crosshatched regions in Figure 6.2 including $\pm L$ (which is the case for the arithmetic given by (6.13)). ∎

We note that condition (6.15) is usually called a *diagonal dominance condition* in the literature [86]. The overflow arithmetic (6.14) has also been considered in [120] where it is called *generalized zeroing arithmetic*. We prefer to use the name generalized overflow arithmetic in our research.

We are now in a position to prove the following result.

Theorem 6.1 The nth order digital filter described by (6.10), in which φ is given in (6.13) or (6.14) with $-1 < L \leq 1$, is free of limit cycles, if A is stable and if there exists a positive definite matrix H which satisfies (6.15), such that

$$Q \triangleq H - A^T H A$$

is positive semidefinite.

We can follow the same procedure as in the proof of Theorem 3.3 to prove that under these conditions, the equilibrium $x_e = 0$ of system (6.10) is globally asymptotically stable. Thus the digital filter described by (6.10) is free of limit cycles.   ∎

For the two's complement and triangular overflow characteristics, we have

Lemma 6.2 An $n \times n$ positive definite matrix $H = [h_{ij}]$ satisfies Assumption (A–6.1) when f represents the two's complement or the triangular arithmetic, *if and only if* H is a diagonal matrix with positive diagonal elements.

The proof is similar to the proof of Lemma 6.1.   ∎

Remark 6.4 A special case of the overflow characteristics given in (6.13) is the zeroing characteristic in which $L = 0$. We can also treat the two's complement and the triangular characteristics as special cases of (6.14) by letting $L = -1$. In this case, condition (6.15) will simply mean that matrix H is a *diagonal positive definite matrix* (which is the result in Lemma 6.2).   ∎

Remark 6.5 For fixed point digital filters employing two's complement or triangular overflow arithmetic, Theorem 6.1 yields the same result as condition (6.3), since for these types of arithmetic, the matrix H which satisfies (A–6.1) must be a *diagonal matrix with positive diagonal elements*. For a digital filter (6.10) using overflow arithmetic given by (6.13) or (6.14) with $-1 < L \leq 1$, our result in Theorem 6.1 relaxes the matrix D in (6.3) from a diagonal matrix with positive diagonal elements to a class of positive definite matrices H which satisfies the condition (6.15). This should certainly cover a broader class of stable matrices A (when $-1 < L \leq 1$).   ∎

For *second-order* digital filters, we have the following Corollary.

Corollary 6.2 Suppose that in a *second-order* digital filter described by (6.10), $A = [a_{ij}]$ is stable and the overflow arithmetic is given by (6.13) or (6.14). A sufficient condition for the non-existence of limit cycles in this digital filter is given by

$$|a_{11} - a_{22}| \leq (1 + L)m + 1 - \det(A), \tag{6.18}$$

if $1 - \det(A) < M - m$, or by

$$|a_{11} - a_{22}| \leq \frac{1 + L}{2}\sqrt{(1 - \det(A))^2 + 4mM}, \tag{6.19}$$

if $1 - \det(A) \geq M - m$, where $M = \max\{|a_{12}|, |a_{21}|\}$ and $m = \min\{|a_{12}|, |a_{21}|\}$.

It is proved in [98] that when (6.9) is satisfied, there exists a 2×2 positive definite matrix H satisfying Assumption (A–3.2) (see page 31), such that $H - A^T H A$ is positive semidefinite, assuming that A is stable. Following the same procedure as in [98], it can be proved that when (6.18) or (6.19) is satisfied, there exists a 2×2 positive definite matrix H satisfying condition (6.15), such that $H - A^T H A$ is positive semidefinite, still assuming that A is stable. ∎

Remark 6.6 Conditions (6.9), (6.18), and (6.19) are applicable only when

$$a_{12}a_{21} < 0,$$

for if A in (6.8) is stable,

$$a_{12}a_{21} \geq 0$$

will guarantee the non-existence of limit cycles in such digital filters for *any* type of overflow nonlinearities satisfying (6.11) (cf. [90] for details). A generalization of condition (6.9) to different types of overflow arithmetic is obtained in [98]. The present result (Corollary 6.2) and the corresponding result given in [98] constitute different conditions which do not cover each other. ∎

6.5 Algorithms for Determining the Matrix H

Theorem 6.1 (resp., Corollary 6.1 (*ii*) or Theorem 3.3) does not specify how to determine a positive definite matrix H which satisfies Assumption (A–6.1) (resp., Assumption (A–3.2)). The existence of such positive definite matrices is sufficient for the global asymptotic stability of the null solution of system (6.10) (resp., (6.1) or (3.11)). In order to apply Theorem 6.1 (resp., Corollary 6.1 (ii) or Theorem 3.3) and to ascertain the global asymptotic stability of the equilibrium $x = 0$ or the non-existence of overflow oscillations for a given fixed-point digital filter with generalized overflow characteristics, it is necessary to determine a positive definite matrix H which satisfies Assumption (A–6.1) (resp., Assumption (A–3.2)). For low order systems, we can usually find matrix H (if it exists) by conducting a search. For high order systems, such an approach is usually impractical. We suggest in the following an algorithm for determining a matrix H for a given coefficient matrix A and the overflow characteristic (which is characterized by the parameter L, $-1 \leq L \leq 1$).

An algorithm for determining matrix H. Suppose in system (6.10) A and L ($-1 \leq L \leq 1$) are given. Consider an objective function given by

$$J = J(H) = \min_i \lambda_i(Q) = \min_i \lambda_i(H - A^T H A) \tag{6.20}$$

where $\lambda_i(Q)$ represents the eigenvalues of the matrix Q and $H = [h_{ij}] \in R^{n \times n}$ satisfy the constraints

$$(1 + L)h_{ii} \geq 2 \sum_{j=1, j \neq i}^{n} |h_{ij}|, \quad i = 1, \cdots, n.$$

If the maximization of the above objective function results in $J > 0$ for a specified coefficient matrix A and the parameter L, we have determined a positive definite matrix H and all conditions of Theorem 6.1 are satisfied. Thus, the null solution of the digital filter (6.10) with such a coefficient matrix and with the generalized overflow nonlinearity given by (6.13) or (6.14) is globally asymptotically stable. ∎

The algorithm proposed above is a *nonlinear* programming problem. To determine a solution for this problem may be quite involved. It turns out that we can modify the above algorithm so that its solution will reduce to a standard *linear* programming problem. The disadvantage introduced by this modification is that a solution to the algorithm is not always guaranteed by the maximization of the *modified* objective function J, given below.

It is well known that a measure of a matrix Q, defined by (cf. Section 3.2)

$$\mu(Q) = \lim_{\theta \to 0+} \frac{\|I + \theta Q\| - 1}{\theta},$$

where $\| \cdot \|$ denotes a matrix norm and I is the identity matrix, serves as an upper bound for the real parts of the eigenvalues of the matrix Q (cf., e.g., [28]). When we consider a symmetric matrix Q, $\mu(Q)$ becomes an upper bound for the eigenvalues of Q. In particular, the relationships between $\mu(Q)$ and $\lambda_i(Q)$ is that

$$Re\lambda_i(Q) \leq \mu(Q),$$

and for a symmetric matrix Q, it is

$$\lambda_i(Q) \leq \mu(Q).$$

We can transform the nonlinear programming problem stated above into a linear programming problem by choosing the objective function to be a measure of the matrix Q, since some of these measures have *linear* relationships with the entries of the matrix. For example, the measures of matrix $Q = [q_{ij}] \in R^{n \times n}$ induced by the matrix norms $\| \cdot \|_1$ and $\| \cdot \|_\infty$ are given by

$$\mu_1(Q) = \max_{1 \leq j \leq n} \left\{ q_{jj} + \sum_{i=1, i \neq j}^{n} |q_{ij}| \right\},$$

and

$$\mu_\infty(Q) = \max_{1 \leq i \leq n} \left\{ q_{ii} + \sum_{j=1, j \neq i}^{n} |q_{ij}| \right\},$$

respectively.

In the present case, since the matrix Q is symmetric, we have

$$\mu_1(Q) = \mu_\infty(Q).$$

Choosing the objective function as

$$J = \mu_1(Q) = \mu_\infty(Q) = \max_{1 \leq i \leq n} \left\{ q_{ii} + \sum_{j=1, j \neq i}^{n} |q_{ij}| \right\}, \tag{6.21}$$

we arrive at a linear programming problem. The maximization of J will sometimes result in a set of large eigenvalues for matrix Q. As mentioned earlier, by choosing an objective function J as in (6.21) and by using the linear programming method to maximize J, we may sometimes not generate a positive definite matrix Q, even if we end up with $J > 0$.

Other alternatives for objective functions, inspired by (6.21), are

$$J = \min_{1 \leq i \leq n} \left\{ q_{ii} - \sum_{j=1, j \neq i}^{n} |q_{ij}| \right\}, \tag{6.22}$$

or

$$J = \min_{1 \leq i \leq n} \left\{ \sigma_i q_{ii} - \sum_{j=1, j \neq i}^{n} \sigma_j |q_{ij}| \right\}, \tag{6.23}$$

where $\sigma_i > 0$ for $i = 1, \cdots, n$. The maximization of J in (6.22) or in (6.23) will always guarantee a set of large eigenvalues, since in our case Q is symmetric. In particular, if the maximization of the objective function J in (6.22) or (6.23) results in $J > 0$, all conditions of Theorem 6.1 are satisfied. (Under these conditions, $Q = Q^T$ becomes a diagonal dominance matrix with positive diagonal elements. Thus, Q is positive definite. See Lemma 6.3 below.) However, the objective function $J > 0$ in (6.22) and (6.23) may yield conservative results, since these are only sufficient conditions for the matrix Q to be positive definite.

Further investigation is expected to determine other objective functions which involve more efficient linear programming problems and whose maximization guarantees the existence of solutions to the problem on hand.

We close the present section with the following lemma.

Lemma 6.3 Assume that $Q = Q^T = [q_{ij}] \in R^{n \times n}$. If there exist $\sigma_i > 0$, $i = 1, \cdots, n$, such that

$$\min_{1 \le i \le n} \left\{ \sigma_i q_{ii} - \sum_{j=1, j \ne i}^{n} \sigma_j |q_{ij}| \right\} > 0, \qquad (6.24)$$

then Q is positive definite.

First, we note that if (6.24) is satisfied, $q_{ii} > 0$ for $i = 1, \cdots, n$. (6.24) then implies that $\overline{Q} = [\overline{q}_{ij}]$ is an M–matrix (cf. [86]), where

$$\overline{q}_{ij} = \begin{cases} q_{ii}, & i = j \\ -|q_{ij}|, & i \ne j \end{cases}.$$

Since $\overline{Q} = \overline{Q}^T$, from the properties of M–matrices [86], $\overline{Q}$ is also positive definite. For any vector $x \in R^n$, we have

$$x^T Q x = \sum_{i=1}^{n} \sum_{j=1}^{n} x_i q_{ij} x_j$$

$$= \sum_{i=1}^{n} q_{ii} x_i^2 + \sum_{i=1}^{n} \sum_{j=1, j \ne i}^{n} x_i q_{ij} x_j$$

$$\ge \sum_{i=1}^{n} q_{ii} x_i^2 - \sum_{i=1}^{n} \sum_{j=1, j \ne i}^{n} |x_i||q_{ij}||x_j|$$

$$= |x^T| \, \overline{Q} \, |x| > 0,$$

since $\overline{Q}$ is positive definite, where $|x| = [|x_1|, \cdots, |x_n|]^T$. Therefore, Q is also positive definite.

∎

6.6 Examples

To demonstrate the applicability of the present results and compare them with existing results, we consider two specific examples.

Example 6.1 For the digital filter (6.1) with

$$A = \begin{bmatrix} 0.6 & -0.2 \\ 0.3 & 1.1 \end{bmatrix},$$ (6.25)

it can be verified that conditions (6.3)–(6.7) are not satisfied.

Choosing H as

$$H = \begin{bmatrix} 1 & 0.5 \\ 0.5 & 0.8 \end{bmatrix},$$ (6.26)

it is easily verified that all conditions of Corollary 6.1 (ii) (or Theorem 3.3) are satisfied. Therefore, the equilibrium $x_e = 0$ of system (6.1) with A specified by (6.25) is globally asymptotically stable. Hence, this digital filter is free of limit cycles. ∎

Condition (6.7) can also be used to ascertain that $x_e = 0$ of the system considered in Example 3.1 on page 33 is globally asymptotically stable. However, application of condition (6.7) is extremely involved and cumbersome. It is extremely difficult to apply condition (6.7) when the order of the system (6.1) (or (3.11)) is greater than 2, as in the next example, where $n = 4$. In particular, the inversion of high-order matrices of variables poses formidable obstacles. On the other hand, the application of Theorem 6.1 to high-order systems is not particularly difficult.

Example 6.2 Consider the system used in Example 3.2 (on page 34) where A is given by

$$A = \begin{bmatrix} -1 & 0 & 0.1 & 0 \\ 0.2 & -0.6 & 0 & 0.8 \\ -0.1 & 0.1 & 0.8 & 0 \\ 0.1 & 0 & 0.1 & -0.5 \end{bmatrix},$$ (6.27)

For this example, it can easily be verified that conditions (6.3)–(6.6) fail as global asymptotic stability tests.

Following the same procedure as in Example 3.2, choosing H as

$$H = \begin{bmatrix} 1.4 & 0 & -0.2 & 0.4 \\ 0 & 1.6 & 0.2 & -0.4 \\ -0.2 & 0.2 & 3.4 & 0.5 \\ 0.4 & -0.4 & 0.5 & 3 \end{bmatrix}, \tag{6.28}$$

we see that the equilibrium $x_e = 0$ of system (6.1) with A specified in (6.27) is globally asymptotically stable. Hence, this digital filter is free of limit cycles.

From Theorem 6.1, we see that a $4th$ order digital filter described by (6.10) with A given in (6.27), when f represents the zeroing arithmetic, is free of limit cycles since H in (6.28) satisfies

$$h_{ii} > 2 \sum_{j=1, j \neq i}^{n} |h_{ij}|, \quad i = 1, \cdots, n.$$

Indeed, it is also free of limit cycles when generalized overflow arithmetic specified in (6.13) or (6.14) is used with $-0.1333 \leq L \leq 1$, from Theorem 6.1. ∎

6.7 Concluding Remarks

The results established in this chapter are new and one of our results (Theorem 6.1) constitutes to the best of our knowledge the least restrictive criterion for testing limit cycles in fixed-point digital filters described by (6.10). Our results are also much easier to apply than existing criteria. The generalized overflow characteristics considered in this chapter (equations (6.13) and (6.14)) cover the usual types of overflow arithmetic used in practice. Our result in Theorem 6.1 for testing whether a digital filter, using the generalized overflow characteristics, is free of limit cycles, constitutes a generalization of condition (6.3), originally given in [51], [90], [114]. We generalize the matrix D in (6.3) from a diagonal matrix with positive diagonal elements to a class of positive definite matrices satisfying condition (6.15). This will certainly cover a larger class of fixed-point digital filters.

Our results can also serve as criteria for testing the global asymptotic stability of digital filters described by (6.10).

CHAPTER 7

STABILITY ANALYSIS OF STATE-SPACE REALIZATIONS FOR
MULTIDIMENSIONAL FILTERS WITH OVERFLOW NONLINEARITIES

7.1 Introduction

As we discussed in Chapter 5, in the *implementation* of linear digital filters, signals are usually represented and processed in a finite wordlength format. Therefore, such implementations will naturally give rise to several kinds of nonlinear effects, such as overflow and quantization. The stability analysis of 2–D (two-dimensional) as well as multidimensional digital filters subject to such nonlinearities has been of increasing interest in recent years (cf. [1], [2], [7], [8], [64], [69], [113]).

Since finite wordlength realizations of digital filters result in systems which inherently are *nonlinear*, the asymptotic stability of such filters (under zero-input), as in many other nonlinear systems, is of great interest in practice. The global asymptotic stability of the null solution guarantees the non-existence of limit cycles (overflow oscillations) in the realized digital filters. In this chapter, we establish new results for the global asymptotic stability of zero-input 2–D state-space digital filters with overflow nonlinearities. We do not consider quantization effects in the present chapter.

The stability properties of 1–D (one-dimensional) digital filters subject to overflow nonlinearities have been investigated extensively during the last two decades (cf. [6], [7], [8], [11], [12], [24], [25], [31], [33], [51], [61], [67], [90], [91], [92], [98], [102], [103], [106], [107], [108], [114], [120], and see also the discussion in Chapter 6). However, a great deal of work which addresses qualitative issues concerning 2–D digital filters endowed with overflow nonlinearities remains to be accomplished.

We consider the state-space quarter plane model of 2–D digital filters described by

$$\begin{bmatrix} x^h(k+1,l) \\ \cdots\cdots\cdots \\ x^v(k,l+1) \end{bmatrix} = f\left(\begin{bmatrix} A_{11} & \vdots & A_{12} \\ \cdots & \vdots & \cdots \\ A_{21} & \vdots & A_{22} \end{bmatrix} \begin{bmatrix} x^h(k,l) \\ \cdots\cdots \\ x^v(k,l) \end{bmatrix} \right), \quad k \geq 0,\ l \geq 0, \qquad (7.1)$$

where $x^h \in R^m$, $x^v \in R^n$, $A_{11} \in R^{m \times m}$, $A_{12} \in R^{m \times n}$, $A_{21} \in R^{n \times m}$, $A_{22} \in R^{n \times n}$, and $f(\cdot)$ represents overflow nonlinearities (which will be defined later in this chapter). We also assume for system (7.1) a finite set of initial conditions, *i.e.*, we assume that there exist two positive integers K and L, such that

$$\begin{cases} x^h(k,0) = 0 \ \text{ for }\ k \geq K, \\ x^v(k,0) = 0 \ \text{ for }\ k \geq K, \\ x^h(0,l) = 0 \ \text{ for }\ l \geq L, \\ x^v(0,l) = 0 \ \text{ for }\ l \geq L. \end{cases} \qquad (7.2)$$

For the asymptotic stability of the 2–D digital filter (7.1) with initial conditions (7.2), a well known result states that (cf. [32], also see [1]) if there exists a diagonal positive definite matrix G such that

$$Q = G - A^T G A \qquad (7.3)$$

is positive definite, then the null solution of the 2–D digital filter (7.1) is globally asymptotically stable, where

$$A = \begin{bmatrix} A_{11} & \vdots & A_{12} \\ \cdots & \vdots & \cdots \\ A_{21} & \vdots & A_{22} \end{bmatrix}. \qquad (7.4)$$

Condition (7.3) can not be applied to the case where the absolute values of some of the diagonal elements of matrix A are greater than or equal to 1.

In this chapter, we utilize Lyapunov's Second Method to establish new results for the global asymptotic stability of the null solution of the 2–D system (7.1). One of our results generalizes condition (7.3) and shows that the matrix G in (7.3) can be relaxed to certain classes of positive definite matrices. We also provide necessary and sufficient conditions under which positive definite matrices can be used to construct for the 2–D system (7.1) quadratic form Lyapunov functions with the desired property that the overflow nonlinearity never increases the values of these functions.

Another result for testing the global asymptotic stability of the null solution for 2–D fixed-point digital filters is given by Bauer and Jury [8] (which is established for the shift-variant case). For shift-invariant 2–D digital filters described by (7.1), the results obtained in [8] require that

$$\rho(|A|) < 1, \tag{7.5}$$

where $\rho(\cdot)$ denotes the spectral radius (see definition on page 42) and $|A| = [|a_{ij}|]$. Condition (7.5) is especially useful for testing the 2-D systems (7.1) with lower or upper triangular coefficient matrices. It can be proved, following the procedure in [2] and [50], that condition (7.5) is a special case of condition (7.3).

We call the class of overflow nonlinearities considered herein *generalized overflow characteristics* (cf. Chapter 6). These nonlinearities constitute a generalization of the usual types of overflow arithmetic employed in practice, including saturation, zeroing, two's complement and triangular overflow arithmetic. In our approach, we do not characterize these nonlinearities by sector conditions; instead, we characterize them by the range of the nonlinear function representing the overflow arithmetic.

In Section 7.2, we prove several results for the global asymptotic stability of the null solution of 2-D digital filters described by (7.1) (Theorem 7.1, Corollary 7.1, and Proposition 7.1). In Section 7.3, we establish our main result for the present chapter (Theorem 7.2). We generalize our results to multidimensional cases in Section 7.4 (Theorem 7.3 and Corollary 7.2). We demonstrate the applicability of the present results by means of specific examples in Section 7.5. We conclude this chapter in Section 7.6.

7.2 General Results for Two-Dimensional Digital Filters

Throughout this chapter, we will use the notation

$$x(k,l) = \begin{bmatrix} x^h(k,l) \\ \cdots\cdots\cdots \\ x^v(k,l) \end{bmatrix} \tag{7.6}$$

and

$$x_{11}(k,l) = \begin{bmatrix} x^h(k+1,l) \\ \cdots\cdots\cdots \\ x^v(k,l+1) \end{bmatrix}, \tag{7.7}$$

for $x^h \in R^m$ and $x^v \in R^n$. Also, we let $D(d)$ denote the set defined by

$$D(d) \triangleq \{(k,l): k+l = d, \ k \geq 0, l \geq 0\}, \tag{7.8}$$

for some positive integer d. (In the context of two-dimensional signal processing, the superscripts h and v suggest the terms "horizontal" and "vertical", respectively, while $D(d)$ suggests indices along a diagonal.)

Consider 2–D shift-invariant systems described by equations of the form

$$\left[\begin{array}{c} x^h(k+1,l) \\ \cdots\cdots\cdots \\ x^v(k,l+1) \end{array}\right] = g\left(\left[\begin{array}{c} x^h(k,l) \\ \cdots\cdots \\ x^v(k,l) \end{array}\right]\right), \quad k \geq 0,\; l \geq 0, \tag{7.9}$$

or compactly,

$$x_{11}(k,l) = g(x(k,l)), \quad k \geq 0,\; l \geq 0, \tag{7.10}$$

where $g\colon R^{m+n} \to R^{m+n}$ is continuous. For such systems, we introduce the following concepts.

Definition 7.1 A point $x_e \in R^{m+n}$ is called an *equilibrium* point of the 2–D system (7.9) (or equivalently, (7.10)) if and only if

$$x_e = g(x_e).$$

Furthermore, if there exists an $r > 0$ such that the open ball

$$B(x_e, r) \triangleq \{x \in R^{m+n} \colon \|x - x_e\| < r\}$$

contains no equilibrium points of (7.9) other than x_e itself, x_e is called an *isolated* equilibrium point, where $\|\cdot\|$ denotes any of the equivalent norms on R^{m+n}. ∎

We assume, without loss of generality, that $x_e = 0$ and that it is isolated. In the following definitions (Definitions 7.2 and 7.3), we assume a finite set of initial conditions as in (7.2) for system (7.9).

Definition 7.2 The equilibrium $x_e = 0$ of the 2–D system (7.9) is said to be *stable* (in the sense of Lyapunov) if for every $\varepsilon > 0$, there exists a $\delta = \delta(\varepsilon) > 0$, such that

$$\|x(k,l)\| < \varepsilon \text{ for all } k \geq 0,\; l \geq 0,$$

whenever

$$\|x(k,0)\| < \delta \text{ for } 0 \leq k \leq K$$

and

$$\|x(0,l)\| < \delta \text{ for } 0 \leq l \leq L$$

where K and L are specified in (7.2). ∎

Definition 7.3 The equilibrium $x_e = 0$ of the 2–D system (7.9) is said to be *globally asymptotically stable* (or *asymptotically stable in the large*) if

(i) it is stable, and

(ii) every solution of (7.9) tends to the origin as $k + l \to \infty$, *i.e.*,

$$\lim_{k \to \infty \text{ and/or } l \to \infty} x(k,l) = \lim_{k+l \to \infty} x(k,l) = 0,$$

for system (7.9) with any initial conditions satisfying (7.2). (Note that in the statement $k + l \to \infty$, we still require that $k \geq 0$ and $l \geq 0$.) In this case, the equilibrium $x_e = 0$ is said to be *globally attractive*. ∎

Remark 7.1 The global asymptotic stability of the equilibrium $x_e = 0$ of (7.9) implies that system (7.9) has *one and only one* equilibrium. ∎

Remark 7.2 In this chapter, we adapt the methodology of the Lyapunov stability theory of an equilibrium for dynamical systems in the qualitative study of 2–D (and multidimensional) filters. In conventional Lyapunov results, a temporal variable (time) plays a central role. In this chapter, which deals with a class of 2–D (and multidimensional) shift-invariant systems, *time* does not have a role. Instead, the independent variables of interest, are *spatial* variables. Therefore, in the present context, asymptotic stability provides the following qualitative characterization of 2–D (and multidimensional) systems:

(i) stability of the equilibrium $x_e = 0$ provides a measure of continuity of the state variables

$$\{x_i(k,l),\ i = 1, \cdots, m + n;\ k = 1, 2, \cdots; l = 1, 2, \cdots\}$$

with respect to a finite set of initial states (see (7.2));

(ii) global attractivity of the origin ensures that the magnitudes of the state variables become arbitrary small when the spatial variables become arbitrary large. ∎

We will employ nonlinearities $f: R^N \to R^N$ to represent overflow effects in 2–D digital filters (7.1), where N denotes the dimension of the underlying vector space,

$$f(x) = [\varphi(x_1), \cdots, \varphi(x_N)]^T, \tag{7.11}$$

and $\varphi: R \to [-1,1]$ is piecewise continuous. We will make use of the notation specified in (7.4) and we will let

$$f(x) = \begin{bmatrix} f(x^h) \\ \cdots\cdots \\ f(x^v) \end{bmatrix}$$

for

$$x = \begin{bmatrix} x^h \\ \cdots \\ x^v \end{bmatrix}.$$

Associated with the nonlinear digital filter (7.1), we will consider linear digital filters given by

$$\begin{bmatrix} w^h(k+1,l) \\ \cdots\cdots\cdots\cdots \\ w^v(k,l+1) \end{bmatrix} = \begin{bmatrix} A_{11} & \vdots & A_{12} \\ \cdots\cdots & \vdots & \cdots\cdots \\ A_{21} & \vdots & A_{22} \end{bmatrix} \begin{bmatrix} w^h(k,l) \\ \cdots\cdots\cdots \\ w^v(k,l) \end{bmatrix}, \quad k \geq 0,\ l \geq 0, \tag{7.12}$$

where A_{11}, A_{12}, A_{21}, and A_{22} are defined in (7.1) and w^h and w^v have compatible dimensions. We will assume a finite set of initial conditions for (7.12) as in (7.2), and follow the convention established in (7.6) and (7.7) for the vector w. The filter model in (7.12) is usually referred to as the Roesser model [99].

In analyzing the stability of the equilibrium $x_e = 0$ of the 2–D system (7.1), we will make use of a class of Lyapunov functions V for the linear system (7.12). In particular, we will make use of the following assumption.

Assumption (A–7.1) Assume that for system (7.12) there exists a continuous function $V: R^{m+n} \to R$ with the following properties:

(i) V can be expressed as

$$V(w) = V^h(w^h) + V^v(w^v) \tag{7.13}$$

where

$$w = \begin{bmatrix} w^h \\ \cdots \\ w^v \end{bmatrix},$$

and $V^h: R^m \to R$ and $V^v: R^n \to R$ are positive definite and radially unbounded. (Thus, V is also positive definite and radially unbounded. See Lemma 7.1 below.) Furthermore, along the solutions of (7.12), V satisfies the condition that

$$DV_{(7.12)}(w(k,l)) \overset{\Delta}{=} V(w_{11}(k,l)) - V(w(k,l))$$

$$= V(Aw(k,l)) - V(w(k,l))$$

is negative definite for all $w(k,l) \in R^{m+n}$ (A is defined in (7.4));

(ii) For all $w \in R^{m+n}$, it is true that

$$V(f(w)) \leq V(w), \tag{7.14}$$

where f represents the overflow nonlinearity for (7.1). ∎

Before establishing our next theorem, we prove the following two preliminary results, Lemmas 7.1 and 7.2.

Lemma 7.1 Assume that $V^h: R^m \to R$ and $V^v: R^n \to R$ are positive definite functions. Define

$$V(w) = V^h(w^h) + V^v(w^v)$$

for

$$w = \begin{bmatrix} w^h \\ \cdots \\ w^v \end{bmatrix},$$

where $w^h \in R^m$ and $w^v \in R^n$. Then, the function $V: R^{m+n} \to R$ is also a positive definite function.

Since V^h and V^v are positive definite, there exist by definition (see Section 2.1) functions $\psi_1 \in \mathcal{K}$ and $\psi_2 \in \mathcal{K}$ such that

$$V^h(w^h) \geq \psi_1(\|w^h\|) \quad \text{for all} \ \|w^h\| \leq r_1$$

and

$$V^v(w^v) \geq \psi_2(\|w^v\|) \quad \text{for all} \ \|w^v\| \leq r_2,$$

for some positive numbers r_1 and r_2. Let

$$\|w\| = \max\{\|w^h\|, \ \|w^v\|\},$$

$$r = \min\{r_1, \ r_2\},$$

and

$$\psi(\|w\|) = \begin{cases} \min\{\psi_1(\|w^h\|), \ \psi_2(\|w^h\|)\} & \text{if} \ \|w^h\| \geq \|w^v\| \\ \min\{\psi_1(\|w^v\|), \ \psi_2(\|w^v\|)\} & \text{if} \ \|w^h\| < \|w^v\| \end{cases}$$

It can easily be shown that ψ is continuous, that $\psi(0) = 0$, and that ψ is strictly increasing on $[0, r]$. Then, $\psi \in \mathcal{K}$, and for any $\|w\| \leq r$

$$V(w) = V^h(w^h) + V^v(w^v)$$

$$\geq \psi_1(\|w^h\|) + \psi_2(\|w^v\|) \geq \psi(\|w\|).$$

Hence, $V \colon R^{m+n} \to R$ is also a positive definite function. ∎

Lemma 7.2 Assume that system (7.1) has a finite set of initial conditions (7.2). For any $\varepsilon > 0$, we can find a $\delta > 0$ such that

$$\max_{0 \leq d \leq \max\{K, L\}} \left\{ \sum_{(k,l) \in D(d)} V(x(k,l)) \right\} < \varepsilon,$$

whenever

$$\|x(k,0)\| < \delta \quad \text{for} \ 0 \leq k \leq K$$

and

$$\|x(0,l)\| < \delta \quad \text{for} \ 0 \leq l \leq L,$$

where the function V is specified in Assumption (A–7.1).

For system (7.12), we define

$$E = \begin{bmatrix} A_{11} & A_{12} \\ 0 & 0 \end{bmatrix}$$

and

$$F = \begin{bmatrix} 0 & 0 \\ A_{21} & A_{22} \end{bmatrix},$$

and we let

$$a = \max\{1, \|E\|_m, \|F\|_m\},$$

where $\| \cdot \|_m$ denotes the matrix norm induced by the vector norm used herein. Thus, (7.12) can be written as

$$w(k+1, l+1) = Ew(k, l+1) + Fw(k+1, l), \quad k \geq 0, l \geq 0. \tag{7.15}$$

Let us consider for (7.12) (or equivalently, (7.15)) a finite set of initial conditions (7.2) with $\|x(k,0)\| < \delta_1$ for $0 \leq k \leq K$ and $\|x(0,l)\| < \delta_1$ for $0 \leq l \leq L$. We now claim that for any integer $d > 0$,

$$\max_{(k,l) \in D(d)} \|w(k,l)\| \leq (2a)^{d-1} \delta_1. \tag{7.16}$$

To prove (7.16), we need only consider $k > 0$ and $l > 0$, since when $k = 0$ or $l = 0$, $\|w(k,l)\| \leq \delta_1$ by assumption. (7.16) is true for $d = 1$, since we need $k = 0$ or $l = 0$ in $(k,l) \in D(1)$. Suppose (7.16) is true for $d = t$, i.e.,

$$\max_{(k,l) \in D(t)} \|w(k,l)\| \leq (2a)^{t-1} \delta_1.$$

Consider now $d = t + 1$, i.e., $(k, l) \in D(t + 1)$ and $\{(k - 1, l), (k, l - 1)\} \subset D(t)$. Since

$$w(k, l) = Ew(k - 1, l) + Fw(k, l - 1),$$

we have for any $(k, l) \in D(t + 1)$, $k > 0$ and $l > 0$,

$$\|w(k, l)\| = \|Ew(k - 1, l) + Fw(k, l - 1)\|$$

$$\leq a(\|w(k - 1, l)\| + \|w(k, l - 1)\|)$$

$$\leq 2a \cdot (2a)^{t-1} \delta_1 = (2a)^t \delta_1.$$

Therefore, when we confine d to $0 \leq d \leq \max\{K, L\}$, we can find a $\delta_1 = \bar{\delta}_1$ small enough such that each component of $w(k, l)$ will never reach the magnitude of 1 for all $(k, l) \in D(d)$, $0 \leq d \leq \max\{K, L\}$. Thus for system (7.1) with a finite set of initial conditions (7.2) and $\|x(k, 0)\| < \bar{\delta}_1$ for $0 \leq k \leq K$ and $\|x(0, l)\| < \bar{\delta}_1$ for $0 \leq l \leq L$, we have $\|x(k, l)\| \leq (2a)^{d-1} \bar{\delta}_1$ and each component of $x(k, l)$ will never reach the magnitude of 1 for all $(k, l) \in D(d)$, $0 \leq d \leq \max\{K, L\}$, since (7.1) is now operating in the linear range. This in turn implies that for the given initial conditions,

$$\max_{(k,l) \in D(d)} \|x(k, l)\| \leq (2a)^{T-2} \bar{\delta}_1,$$

for all $d \in [0, \max\{K, L\}]$, where $T = \max\{K, L\} + 1$.

For the given $\varepsilon > 0$, we can find a $\delta_2 > 0$ such that $V(x) < \varepsilon/T$ whenever $\|x\| < \delta_2$. Choose

$$\delta = \min\left\{\bar{\delta}_1, \frac{\delta_2}{(2a)^{T-2}}\right\}.$$

Then, $\|x(k, 0)\| < \delta$ for $0 \leq k \leq K$ and $\|x(0, l)\| < \delta$ for $0 \leq l \leq L$ imply that

$$\max_{0 \leq d \leq \max\{K,L\}} \left\{\max_{(k,l) \in D(d)} \|x(k, l)\|\right\} \leq (2a)^{T-2} \delta \leq \delta_2.$$

This in turn implies that

$$\max_{0 \leq d \leq \max\{K,L\}} \left\{\max_{(k,l) \in D(d)} V(x(k, l))\right\} < \frac{\varepsilon}{T}.$$

Therefore,

$$\max_{0 \leq d \leq \max\{K,L\}} \left\{\sum_{(k,l) \in D(d)} V(x(k, l))\right\} \leq \max_{0 \leq d \leq \max\{K,L\}} \left\{(d + 1) \max_{(k,l) \in D(d)} V(x(k, l))\right\}$$

$$\leq \max_{0 \leq d \leq \max\{K,L\}} \left\{T \max_{(k,l) \in D(d)} V(x(k, l))\right\} < T \cdot \frac{\varepsilon}{T} = \varepsilon.$$

This proves the lemma. ∎

We are now in a position to establish the following result.

Theorem 7.1 If Assumption (A–7.1) holds, the equilibrium $x_e = 0$ of the 2–D system (7.1) is globally asymptotically stable.

Since (A–7.1) is true, there exist positive definite and radially unbounded functions V, V^h, and V^v for system (7.12), such that (7.14) is true, which in turn implies that

$$V(f(Aw)) \leq V(Aw) \text{ for all } w \in R^{m+n}.$$

Also, by (A–7.1),

$$V(Aw(k,l)) < V(w(k,l)) \text{ for all } w(k,l) \neq 0.$$

Thus, for the 2–D system (7.1), we have, using (7.13) and (7.14),

$$V(x_{11}(k,l)) = V(f(Ax(k,l))$$

$$\leq V(Ax(k,l)) < V(x(k,l)) \text{ for all } x(k,l) \neq 0, \tag{7.17}$$

i.e.,

$$V(x_{11}(k,l)) = V^h(x^h(k+1,l)) + V^v(x^v(k,l+1)) < V(x(k,l)) \tag{7.18}$$

for all $x(k,l) \neq 0$.

For any integer $d \geq \max\{K, L\}$, we compute

$$\sum_{(k,l)\in D(d)} V(x(k,l)) > \sum_{(k,l)\in D(d)} V(x_{11}(k,l))$$

$$= \sum_{(k,l)\in D(d)} [V^h(x^h(k+1,l)) + V^v(x^v(k,l+1))]$$

$$= \sum_{(k,l)\in D(d)} V^h(x^h(k+1,l)) + V^h(x^h(0,d+1))$$

$$+ \sum_{(k,l)\in D(d)} V^v(x^v(k,l+1)) + V^v(x^v(d+1,0))$$

$$= \sum_{(k,l)\in D(d+1)} V^h(x^h(k,l)) + \sum_{(k,l)\in D(d+1)} V^v(x^v(k,l))$$

$$= \sum_{(k,l)\in D(d+1)} V(x(k,l)). \tag{7.19}$$

In the above, we have used the fact that

$$x^h(0,d+1) = 0,$$

$$x^v(d+1,0) = 0,$$

and the positive definiteness of the functions V^h and V^v.

Consider any fixed $\varepsilon > 0$. Since V is radially unbounded, there exists a function $\psi_1 \in \mathcal{KR}$ such that $V(0) = 0$ and $V(x) \geq \psi_1(\|x\|)$ for all x satisfying $\|x\| < \varepsilon + 1$. Pick $\delta > 0$ so small that

$$\max_{0 \leq d \leq \max\{K,L\}} \left\{ \sum_{(k,l) \in D(d)} V(x(k,l)) \right\} < \psi_1(\varepsilon), \qquad (7.20)$$

whenever

$$\|x(k,0)\| < \delta \text{ for } 0 \leq k \leq K$$

and

$$\|x(0,l)\| < \delta \text{ for } 0 \leq l \leq L.$$

This is always possible since K and L are finite (see Lemma 7.2 above). Then (7.19) and (7.20) imply that

$$\sum_{(k,l) \in D(d)} V(x(k,l)) < \psi_1(\varepsilon) \text{ for all } d \geq 0. \qquad (7.21)$$

Hence, $\|x(k,l)\|$ can not reach the value ε for all $k \geq 0$ and $l \geq 0$, since this would imply that

$$V(x(k,l)) \geq \psi_1(\|x(k,l)\|) = \psi_1(\varepsilon), \qquad (7.22)$$

which contradicts (7.21). Therefore, the equilibrium $x_e = 0$ of the 2-D system (7.1) is stable (see definition 7.2).

To complete the proof of the theorem, we must show that for any initial conditions satisfying (7.2),

$$\lim_{\substack{k \to \infty \text{ and/or } l \to \infty}} x(k,l) = \lim_{k+l \to \infty} x(k,l) = 0.$$

Since we have overflow nonlinearities in (7.1), we may assume that $\|x(k,l)\| < C$ for all $k \geq 0$ and $l \geq 0$ for some finite real number $C > 0$, without loss of generality. We now define

$$DV_{(7.1)}(x(k,l)) \triangleq V(x_{11}(k,l)) - V(x(k,l))$$

$$= V(f(Ax(k,l))) - V(x(k,l)).$$

(7.17) implies that $DV_{(7.1)}(x(k,l))$ is negative definite for *all* $x(k,l) \in R^{m+n}$. Hence, there exists a function $\psi_2 \in \mathcal{K}$ such that $DV_{(7.1)}(0) = 0$ and $DV_{(7.1)}(x) \leq -\psi_2(\|x\|)$ for all x satisfying $\|x\| < C$. Following the same argument as in (7.19), we now have, for any

$d \geq \max\{K, L\}$,

$$\sum_{(k,l)\in D(d)} DV_{(7.1)}(x(k,l)) = \sum_{(k,l)\in D(d)} V(x_{11}(k,l)) - \sum_{(k,l)\in D(d)} V(x(k,l))$$

$$= \sum_{(k,l)\in D(d+1)} V(x(k,l)) - \sum_{(k,l)\in D(d)} V(x(k,l))$$

$$\leq - \sum_{(k,l)\in D(d)} \psi_2(\|x(k,l)\|). \qquad (7.23)$$

Since V is positive definite and radially unbounded and $\psi_2 \in \mathcal{K}$, relation (7.23) implies that for (7.1) with any initial conditions satisfying (7.2),

$$\lim_{d\to\infty}\left[\sum_{(k,l)\in D(d+1)} V(x(k,l)) - \sum_{(k,l)\in D(d)} V(x(k,l)) \right] = 0.$$

This in turn implies that

$$\lim_{d\to\infty} \sum_{(k,l)\in D(d)} \psi_2(\|x(k,l)\|) = 0.$$

It follows that

$$\psi_2(\|x(k,l)\|) \to 0 \text{ as } k + l \to \infty.$$

Therefore, for (7.1) with any initial conditions satisfying (7.2), we have that

$$x(k,l) \to 0 \text{ as } k \to \infty \text{ and/or } l \to \infty. \qquad \blacksquare$$

We will refer to a V function satisfying Theorem 7.1 as a *Lyapunov function* for the 2–D system (7.1).

In particular, when we choose the function V as the pth power of the l_p vector norm, $1 \leq p < \infty$,

$$V(w) = \|w\|_p^p = \sum_{i=1}^{m+n} |w_i|^p, \qquad (7.24)$$

we have the following result.

Corollary 7.1 The equilibrium $x_e = 0$ of the 2–D system (7.1) is globally asymptotically stable if

$$\|A\|_p < 1, \text{ for some } p, \ 1 \leq p < \infty, \qquad (7.25)$$

where $\| \cdot \|_p$ denotes the matrix norm induced by the l_p vector norm.

It suffices to show that if (7.25) is true then Assumption (A–7.1) is satisfied.

Clearly,

$$V(w) = \|w\|_p^p = \sum_{i=1}^{m+n} |w_i|^p = \sum_{i=1}^{m} |w_i^h|^p + \sum_{i=1}^{n} |w_i^v|^p$$
$$= \|w^h\|_p^p + \|w^v\|_p^p = V^h(w^h) + V^v(w^v),$$

where V^h and V^v are defined in the obvious way and are positive definite and radially unbounded. Also, in view of (7.25), we have

$$V(w_{11}(k,l)) = \|Aw(k,l)\|_p^p \leq \|A\|_p^p \|w(k,l)\|_p^p$$

$$< \|w(k,l)\|_p^p = V(w(k,l)),$$

for all $w(k,l) \neq 0$. Thus, Assumption (A–7.1) (i) is satisfied. Assumption (A–7.1) (ii) is also satisfied since

$$\|f(w)\|_p \leq \|w\|_p \tag{7.26}$$

holds for any p, $1 \leq p \leq \infty$, and *any* type of *overflow* nonlinearity given in (7.11). ∎

Remark 7.3 For 1–D fixed-point digital filters given by

$$x(k+1) = f(Ax(k)), \ k \geq 0, \tag{7.27}$$

condition $\|A\|_p < 1$ for some p, $1 \leq p \leq \infty$, guarantees the global asymptotic stability of the null solution of the digital filter (7.27) (cf. Remark 3.4 of Section 3.5). For 2–D digital filters described by (7.1), we proved in the above corollary that condition (7.25) (where $1 \leq p < \infty$) guarantees the global asymptotic stability of the equilibrium $x = 0$ for such filters. A special case of condition (7.25),

$$\|A\|_2 < 1, \tag{7.28}$$

has been proved in [1] using a slightly different method. (In [1], (7.28) was considered as a special case of condition (7.3).) ∎

Next, we prove that

$$\|A\|_\infty < 1 \tag{7.29}$$

is also a sufficient condition for the global asymptotic stability of the null solution of the 2–D system (7.1), using a different approach from the one used above. Condition (7.29) does not appear to be readily obtainable from Theorem 7.1.

Proposition 7.1 Condition (7.29) is a sufficient condition for the global asymptotic stability of the equilibrium $x_e = 0$ of the 2–D systems (7.1).

Choose a function V for system (7.1) as

$$V(x) = \|x\|_\infty \overset{\triangle}{=} \max_{1 \le i \le m+n} \{|x_i|\}.$$

By the definition of the l_∞ vector norm and in view of (7.26) and (7.29), we have

$$V(x) = \max\{V^h(x^h), V^v(x^v)\}, \tag{7.30}$$

and

$$V(x_{11}(k,l)) = V(f(Ax(k,l))$$

$$= \|f(Ax(k,l))\|_\infty \le \|Ax(k,l)\|_\infty$$

$$\le \|A\|_\infty \|x(k,l)\|_\infty < \|x(k,l)\|_\infty = V(x(k,l)), \tag{7.31}$$

for all $x(k,l) \ne 0$, where $V^h(x^h) = \|x^h\|_\infty$ and $V^v(x^v) = \|x^v\|_\infty$. Relation (7.31) can be rewritten as

$$V(x_{11}(k,l)) = \max\{V^h(x^h(k+1,l)), V^v(x^v(k,l+1))\}$$

$$< V(x(k,l)), \tag{7.32}$$

for all $x(k,l) \ne 0$, $k \ge 0$, $l \ge 0$.

For $(k,l) \in D(d)$, $d \ge \max\{K,L\}$, let

$$\max_{(k,l) \in D(d)} \{V(x(k,l))\} = V(x(p_d, q_d)), \tag{7.33}$$

where $p_d + q_d = d$. From (7.32), we see that

$$V^h(x^h(p_{d+1}, q_{d+1})) \le \max\{V^h(x^h(p_{d+1}, q_{d+1})), V^v(x^v(p_{d+1} - 1, q_{d+1} + 1))\}$$

$$< V(x(p_{d+1} - 1, q_{d+1})) \tag{7.34}$$

and

$$V^v(x^v(p_{d+1}, q_{d+1})) \le \max\{V^h(x^h(p_{d+1} + 1, q_{d+1} - 1)), V^v(x^v(p_{d+1}, q_{d+1}))\}$$

$$< V(x(p_{d+1}, q_{d+1} - 1)). \tag{7.35}$$

Clearly, from (7.30) and (7.33)–(7.35) and from the fact that

$$\{(p_{d+1} - 1, q_{d+1}), (p_{d+1}, q_{d+1} - 1)\} \subset D(d),$$

we now have

$$\max_{(k,l) \in D(d+1)} \{V(x(k,l))\} = V(x(p_{d+1}, q_{d+1}))$$

$$= \max\{V^h(x^h(p_{d+1}, q_{d+1})), V^v(x^v(p_{d+1}, q_{d+1}))\}$$

$$< \max\{V(x(p_{d+1} - 1, q_{d+1})), V(x(p_{d+1}, q_{d+1} - 1))\}$$

$$\leq \max_{(k,l) \in D(d)} \{V(x(k,l))\} = V(x(p_d, q_d)), \qquad (7.36)$$

when $x(k,l) \neq 0$ for some $(k,l) \in D(d)$, $d \geq \max\{K, L\}$.

From (7.36), the proof of stability of the equilibrium $x_e = 0$ of (7.1) follows along similar lines as the proof of stability in Theorem 7.1.

We now prove that the equilibrium $x_e = 0$ of (7.1) is globally attractive, $i.e.$, for system (7.1) with any initial conditions satisfying (7.2), $x(k,l) \to 0$ as $k + l \to \infty$ with $k \geq 0$ and $l \geq 0$.

Relation (7.31) implies that

$$DV_{(7.1)}(x(k,l)) \triangleq V(x_{11}(k,l)) - V(x(k,l))$$

$$= V(f(Ax(k,l))) - V(x(k,l))$$

is negative definite for all $x(k,l) \in R^{m+n}$. Hence, there exists a function $\psi \in \mathcal{K}$ such that $DV_{(7.1)}(0) = 0$ and $DV_{(7.1)}(x) \leq -\psi(\|x\|)$ for all $\|x\| < C$, $i.e.$,

$$V(x_{11}(k,l)) - V(x(k,l)) = \max\{V^h(x^h(k+1,l)), V^v(x^v(k,l+1))\} - V(x(k,l))$$

$$\leq -\psi(\|x(k,l)\|). \qquad (7.37)$$

(C is specified in the proof of Theorem 7.1.)

Denote

$$V^s(x^s(p_{d+1}, q_{d+1})) \triangleq \max\{V^h(x^h(p_{d+1}, q_{d+1})), V^v(x^v(p_{d+1}, q_{d+1}))\}$$

$$= V(x(p_{d+1}, q_{d+1})) \qquad (7.38)$$

where

$$s = \begin{cases} h, & if \ V^h(x^h(p_{d+1}, q_{d+1})) \geq V^v(x^v(p_{d+1}, q_{d+1})) \\ v, & if \ V^h(x^h(p_{d+1}, q_{d+1})) < V^v(x^v(p_{d+1}, q_{d+1})) \end{cases}.$$

Relations (7.34) and (7.35) can be written as

$$V^s(x^s(p_{d+1}, q_{d+1})) < V(x(p_{ds}, q_{ds})), \qquad (7.39)$$

where

$$p_{ds} = \begin{cases} p_{d+1} - 1, & \text{if } s = h \\ p_{d+1}, & \text{if } s = v \end{cases},$$

and

$$q_{ds} = \begin{cases} q_{d+1}, & \text{if } s = h \\ q_{d+1} - 1, & \text{if } s = v \end{cases}.$$

Now using (7.33), (7.37) and (7.38) and with $(p_{ds}, q_{ds}) \in D(d)$ defined above, we have, when $s = h$,

$$V(x(p_{d+1}, q_{d+1})) - V(x(p_d, q_d)) \leq V^h(x^h(p_{ds} + 1, q_{ds})) - V(x(p_{ds}, q_{ds}))$$

$$\leq -\psi(\|x(p_{ds}, q_{ds})\|), \tag{7.40}$$

and when $s = v$,

$$V(x(p_{d+1}, q_{d+1})) - V(x(p_d, q_d)) \leq V^v(x^v(p_{ds}, q_{ds} + 1)) - V(x(p_{ds}, q_{ds}))$$

$$\leq -\psi(\|x(p_{ds}, q_{ds})\|). \tag{7.41}$$

Since V is positive definite and radially unbounded, (7.40) and (7.41) imply that

$$\lim_{d \to \infty} V(x(p_d, q_d)) = r, \tag{7.42}$$

for system (7.1) with any initial conditions satisfying (7.2), where $r \geq 0$. We next prove that r in (7.42) is in fact zero.

Relations (7.40)–(7.42) $\Rightarrow$

$$\lim_{d \to \infty} \psi(\|x(p_{ds}, q_{ds})\|) = 0.$$

$\Rightarrow$

$$\lim_{d \to \infty} x(p_{ds}, q_{ds}) = 0.$$

$\Rightarrow$

$$\lim_{d \to \infty} V(x(p_{ds}, q_{ds})) = 0.$$

Considering (7.38) and (7.39), this will in turn imply that

$$\lim_{d \to \infty} V(x(p_{d+1}, q_{d+1})) = \lim_{d \to \infty} V^s(x^s(p_{d+1}, q_{d+1}))$$

$$\leq \lim_{d \to \infty} V(x(p_{ds}, q_{ds})) = 0,$$

or equivalently,

$$\lim_{d \to \infty} V(x(p_d, q_d)) = 0.$$

Thus, $V(x(k,l)) \to 0$ as $d \to \infty$ for all $(k,l) \in D(d)$, since

$$V(x(p_d, q_d)) = \max_{(k,l) \in D(d)} \{V(x(k,l))\}.$$

Therefore $x(k,l) \to 0$ as $k + l \to \infty$.

It now follows that the equilibrium $x_e = 0$ of the 2–D system (7.1) is globally asymptotically stable. ∎

Remark 7.4 Summarizing the results in Corollary 7.1 and Proposition 7.1, we can now conclude that the equilibrium $x_e = 0$ of the 2–D system (7.1) is globally asymptotically stable if

$$\|A\|_p < 1, \text{ for some } p, \ 1 \le p \le \infty. \tag{7.43}$$

We note that condition (7.43) constitutes also a criterion for the global asymptotic stability of the equilibrium $w_e = 0$ of the linear 2–D system described by (7.12). ∎

7.3 Main Result for Two-Dimensional Digital Filters

In the sequel, we utilize the *generalized overflow arithmetic* introduced in Section 6.4 where the function φ in (7.11) is given by

$$\varphi(x_i) = \begin{cases} L, & x_i > 1 \\ x_i, & -1 \le x_i \le 1 \\ -L, & x_i < -1 \end{cases} \tag{7.44}$$

or

$$\begin{cases} L \le \varphi(x_i) \le 1, & x_i > 1 \\ \varphi(x_i) = x_i, & -1 \le x_i \le 1 \\ -1 \le \varphi(x_i) \le -L, & x_i < -1 \end{cases} \tag{7.45}$$

where $-1 \le L \le 1$.

In the following, we will consider quadratic form Lyapunov functions for system (7.1). In deriving our next result, we make use of the following assumption.

Assumption (A– 7.2) Let f be defined as in (7.11). Assume that there exists a positive definite matrix $H \in R^{N \times N}$ such that

$$f(x)^T H f(x) < x^T H x$$

for all $x \in R^N$, $x \notin D^N \triangleq \{x \in R^N: -1 \le x_i \le 1\}$. ∎

Our next result provides a *necessary and sufficient* condition for matrices to satisfy Assumption (A–7.2) when f represents the generalized overflow arithmetic.

Lemma 7.1 Assume that f is defined in (7.11) and φ is defined in (7.44) or (7.45), where $-1 \leq L \leq 1$. An $N \times N$ positive definite matrix $H = [h_{ij}]$ satisfies Assumption (A–7.2) *if and only if*

$$(1 + L)h_{ii} \geq 2 \sum_{j=1,j\neq i}^{N} |h_{ij}|, \quad i = 1, \cdots, N. \tag{7.46}$$

This lemma is a combined version of Lemmas 6.1 and 6.2 in Section 6.4. ∎

We single out the following special cases of the above lemma:

1. When in (7.44), $L = 1$, f represents the *saturation overflow nonlinearity* and (7.46) assumes the form

$$h_{ii} \geq \sum_{j=1,j\neq i}^{N} |h_{ij}|, \quad i = 1, \cdots, N.$$

This case represents a *diagonal dominance condition*.

2. When in (7.44), $L = 0$, f represents the *zeroing arithmetic* and (7.46) assumes the form

$$h_{ii} \geq 2 \sum_{j=1,j\neq i}^{N} |h_{ij}|, \quad i = 1, \cdots, N. \tag{7.47}$$

3. When in (7.45), $L = -1$, f represents overflow nonlinearities including the *two's complement arithmetic* and *triangular arithmetic*. For such cases, (7.46) assumes the form

$$h_{ij} = 0, \quad j \neq i, \ j = 1, \cdots, N,$$

i.e., H is a diagonal matrix with positive diagonal elements.

We are now in a position to prove the following result.

Theorem 7.2 The equilibrium $x_e = 0$ of the 2–D digital filter (7.1) is globally asymptotically stable, *if* there exist positive definite matrices $H^h \in R^{m \times m}$ and $H^v \in R^{n \times n}$ satisfying Assumption (A–7.2) (with $N = m$ and $N = n$, respectively), such that

$$Q = H - A^T H A \tag{7.48}$$

is positive definite, where

$$H = H^h \oplus H^v \triangleq \begin{bmatrix} H^h & \vdots & 0 \\ \cdots & \vdots & \cdots \\ 0 & \vdots & H^v \end{bmatrix}.$$

For (7.1) we choose the positive definite and radially unbounded Lyapunov function $V(x) = x^T H x$. Since H^h and H^v satisfy Assumption (A-7.2), we have

$$V(f(x)) = f(x)^T H f(x)$$

$$= f(x^h)^T H^h f(x^h) + f(x^v)^T H^v f(x^v)$$

$$\leq (x^h)^T H^h x^h + (x^v)^T H^v x^v = V(x)$$

for all $x \in R^{m+n}$. Thus,

$$V(x_{11}(k,l)) = f(Ax(k,l))^T H f(Ax(k,l))$$

$$\leq x(k,l)^T A^T H A x(k,l) \tag{7.49}$$

for all $x(k,l) \in R^{m+n}$. We now have

$$V(x_{11}(k,l)) \leq x(k,l)^T A^T H A x(k,l)$$

$$< x(k,l)^T H x(k,l) = V(x(k,l)) \tag{7.50}$$

for all $x(k,l) \neq 0$, since $H - A^T H A$ is positive definite.

The rest of the proof follows along similar lines as the proof of Theorem 7.1. ∎

Remark 7.5 Theorem 7.2 constitutes a generalization of condition (7.3). Specifically, we relax the matrix G in (7.3) from a diagonal positive definite matrix to a class of positive definite matrices H which is generated from two positive definite matrices satisfying condition (7.46). This should certainly cover a broader class of coefficient matrices A for 2-D digital filters described by (7.1) using the generalized overflow nonlinearity with $L > -1$ than condition (7.3). ∎

Remark 7.6 In Section 7.5, we consider a specific example (Example 7.1) which suggests that results provided in Theorem 7.2 are less conservative than the conditions (7.25) and (7.29) (or equivalently, condition (7.43)). Indeed, this example can be analyzed by Theorem 7.2, but not by conditions (7.25) and (7.29). ∎

Remark 7.7 In a result which corresponds to Theorem 7.2 for 1–D digital filters described by (7.27), we only require that in $Q = H - A^T H A$, matrix H satisfy Assumption (A–7.2) and that matrix Q be positive semidefinite (under the assumption that A is stable, cf. Section 6.4). A further similar relaxation for the matrix Q in condition (7.48) has not been achieved thus far. ∎

Remark 7.8 Theorem 7.2 yields sufficient conditions for the global asymptotic stability of the equilibrium $x_e = 0$ of the 2–D system (7.1). Its applications require the existence of two positive definite matrices H^h and H^v which satisfy condition (7.46). The algorithm which we developed in Section 6.5 can easily be modified so that we can determine matrices H^h and H^v for a given matrix A and a number L $(-1 \leq L \leq 1)$ using linear programming. ∎

Remark 7.9 The results developed in this section can be used directly as criteria for non-existence of limit cycles (under zero input conditions) of 2–D digital filters described by (7.1). Recently, Tzafestas *et al* reported some new conditions for non-existence of overflow oscillations of 2–D digital filters subject to overflow nonlinearities [113]. Our results in this section are more general than the results obtained in [113] since we consider the global asymptotic stability of the equilibrium $x_e = 0$ of 2–D digital filters subject to overflow nonlinearities and since the results in [113] require that *either* A_{11} (or A_{22}) be a scalar *or* the overflow nonlinearity f satisfy

$$f(x)^T E [x - f(x)] \geq 0 \ \text{ for all } \ x \in R^{m+n},$$

where E is a positive definite diagonal matrix. ∎

7.4 Multidimensional Digital Filters with Overflow Nonlinearities

In this section, we consider m–D (multidimensional or m-dimensional) digital filters described by equations of the form

$$x_I(k_1, \cdots, k_m) = f(Ax(k_1, \cdots, k_m)), \tag{7.51}$$

where $k_i \geq 0$ for $i = 1, \cdots, m$,

$$x(k_1, \cdots, k_m) = \begin{bmatrix} x^1(k_1, \cdots, k_m) \\ \vdots \\ x^m(k_1, \cdots, k_m) \end{bmatrix},$$

97

$x^i \in R^{n_i}$ for $i = 1, \cdots, m$,

$$x_I(k_1, k_2, \cdots, k_m) = \begin{bmatrix} x^1(k_1 + 1, k_2, \cdots, k_m) \\ x^2(k_1, k_2 + 1, \cdots, k_m) \\ \vdots \\ x^m(k_1, k_2, \cdots, k_m + 1) \end{bmatrix},$$

A has compatible dimension and structure, and f represents the overflow nonlinearities defined in (7.11). We assume for system (7.51) a finite set of initial conditions, $i.e.$, we assume that for $i = 1, \cdots, m$,

$$x^i(k_1, 0, \cdots, 0) = 0 \quad \text{for} \quad k_1 \geq K_1,$$

$$x^i(0, k_2, \cdots, 0) = 0 \quad \text{for} \quad k_2 \geq K_2,$$

$$\vdots$$

$$x^i(0, 0, \cdots, k_m) = 0 \quad \text{for} \quad k_m \geq K_m,$$

where $K_1, \cdots, K_m$ are finite positive integers.

The definitions for the stability of 2–D systems given in Section 7.2 can be generalized to m–D systems in the obvious way. Furthermore, Theorems 7.1 and 7.2, Corollary 7.1, and Proposition 7.1 (applicable to 2–D systems) can also be generalized to m–D systems in the obvious way. The following results constitute generalizations of Theorem 7.2, Corollary 7.1 and Proposition 7.1 to m–D systems described by (7.51). Their proofs follow along similar lines as the proofs of the corresponding results for the 2–D case.

Theorem 7.3 The equilibrium $x_e = 0$ of the m–D digital filter (7.51) is globally asymptotically stable, if there exist positive definite matrices $H^i \in R^{n_i \times n_i}$ for $i = 1, \cdots, m$, satisfying Assumption (A–7.2) (with $N = n_i$, $i = 1, \cdots, m$, respectively), such that

$$Q = H - A^T H A \tag{7.52}$$

is positive definite, where

$$H = H^1 \oplus H^2 \oplus \cdots \oplus H^m \triangleq \begin{bmatrix} H^1 & 0 & \cdots & 0 \\ 0 & H^2 & \cdots & 0 \\ \vdots & \vdots & \ddots & \vdots \\ 0 & 0 & \cdots & H^m \end{bmatrix}.$$

Corollary 7.2 The equilibrium $x_e = 0$ of the m–D system (7.51) is globally asymptotically stable *if*

$$\|A\|_p < 1, \text{ for some } p, \ 1 \le p \le \infty. \tag{7.53}$$

∎

Remark 7.10 Theorem 7.1 can also be generalized to m–D systems in a straightforward manner. In generalizing Theorem 7.1, we define (analogous to $D(d)$ given in (7.8)) the hyperplane denoted by $D(M)$,

$$D(M) \triangleq \{(k_1, \cdots, k_m): k_1 + \cdots + k_m = M, \ k_i \ge 0, \ i = 1, \cdots, m\}, \tag{7.54}$$

for some integer $M > 0$. The generalization of Theorem 7.1 (Assumption (A–7.1)) to m–D systems involves the existence of a Lyapunov function $V: R^P \to R$, where $P = n_1 + \cdots + n_m$, with the following properties:

(i) V can be expressed as a sum of functions $V^i: R^{n_i} \to R$, $i = 1, \cdots, m$,

$$V(x) = V^1(x^1) + \cdots + V^m(x^m),$$

(ii) each V^i is a function of the partial state x^i only,

(iii) every V^i is positive definite and radially unbounded (in the state x^i),

(iv) the function V satisfies that

$$DV_{(L)}(x(k_1, \cdots, k_m)) \triangleq V(Ax(k_1, \cdots, k_m)) - V(x(k_1, \cdots, k_m))$$

is negative definite for all $x(k_1, \cdots, k_m) \in R^P$, and

(v) for all $x \in R^P$, it is true that

$$V(f(x)) \le V(x),$$

where f represents the overflow nonlinearities for system (7.51).

It should be noted that for system (7.51), if (7.53) is satisfied for some p, $1 \le p < \infty$, conditions (i)–(v) above will be satisfied by choosing V as in (7.24). ∎

7.5 Examples

To demonstrate the applicability of the present results and compare them with existing results, we now consider two specific examples: a two-dimensional digital filter, and a three-dimensional digital filter.

Example 7.1 It can easily be verified that for a $2h$–$2v$ 2-D digital filter (7.1) using saturation arithmetic with A given by

$$A = \left[\begin{array}{cc:cc} -1 & -0.2 & 0.1 & -0.2 \\ 0.2 & -0.6 & 0 & 0.6 \\ \hdashline -0.1 & 0.1 & 0.5 & -0.9 \\ 0.1 & 0 & 0.1 & -0.5 \end{array}\right], \tag{7.55}$$

we have $\|A\|_p > 1$, for $p = 1, 2, \infty$, and there is no diagonal matrix G with positive diagonal elements such that $G - A^T G A$ is positive definite. Hence, conditions (7.3), (7.25) (for $p = 1, 2$), and (7.29) fail as global asymptotic stability tests for the present example.

Hypothesis (A–7.2) is satisfied for this example by choosing

$$H^h = \begin{bmatrix} 0.9 & 0.3 \\ 0.3 & 0.9 \end{bmatrix}$$

and

$$H^v = \begin{bmatrix} 0.6 & 0.2 \\ 0.2 & 6.4 \end{bmatrix}.$$

Then,

$$H = H^h \oplus H^v = \left[\begin{array}{cc:cc} 0.9 & 0.3 & 0 & 0 \\ 0.3 & 0.9 & 0 & 0 \\ \hdashline 0 & 0 & 0.6 & 0.2 \\ 0 & 0 & 0.2 & 6.4 \end{array}\right]. \tag{7.56}$$

Since

$$Q = H - A^T H A = \begin{bmatrix} 0.018 & 0.064 & 0.042 & 0.178 \\ 0.064 & 0.462 & 0.004 & 0.352 \\ 0.042 & 0.004 & 0.357 & 0.858 \\ 0.178 & 0.352 & 0.858 & 3.846 \end{bmatrix}$$

is positive definite, all conditions of Theorem 7.2 are satisfied, and the equilibrium $x_e = 0$ of the 2–D system (7.1) using saturation arithmetic with coefficient matrix given in (7.55) is globally asymptotically stable.

When f represents the *zeroing* arithmetic, we see from Theorem 7.2 that the equilibrium $x_e = 0$ of the 2–D digital filter described by (7.1) with A given in (7.55) is globally asymptotically stable since H^h and H^v satisfy (cf (7.47))

$$h_{ii} > 2 \sum_{j=1, j \neq i}^{2} |h_{ij}|, \quad i = 1, 2.$$

Indeed, the equilibrium $x_e = 0$ of (7.1) with A given in (7.55) is also globally asymptotically stable when *generalized overflow arithmetic* specified by (7.44) or (7.45) is used with $-0.3333 \leq L \leq 1$. ∎

Example 7.2 Consider a 3–D (2–2–3) digital filter described by (7.51) with *saturation* overflow arithmetic and with A given by

$$A = \left[\begin{array}{cc:cc:ccc} 0.8 & -0.2 & 0 & -0.2 & 0 & -0.2 & 0 \\ -0.4 & -0.5 & 0 & -0.2 & 0.2 & 0 & 0 \\ \hdashline 0.1 & 0.25 & 0.5 & -0.5 & 0.4 & -0.2 & -0.1 \\ 0 & 0 & 0.1 & -0.5 & 0.05 & 0.1 & 0 \\ \hdashline 0 & -0.3 & 0 & 0.1 & -0.2 & 0.54 & 0.4 \\ 0.15 & 0 & 0 & 0 & -0.1 & 1 & 0.1 \\ 0 & 0.08 & -0.05 & 0 & 0.05 & 0.3 & -0.8 \end{array} \right]. \quad (7.57)$$

It can be verified that $\|A\|_p > 1$ for $p = 1, 2$, and ∞. Also, it can be verified that there is no diagonal matrix G with positive diagonal elements such that $G - A^T G A$ is positive definite. (It is clear that condition (7.3) can be applied to the m–D case.) Hence, we attempt to apply Theorem 7.3 for this example.

According to (7.46) in Lemma 7.1 for $L = 1$, we can choose

$$H^1 = \begin{bmatrix} 0.7 & 0.2 \\ 0.2 & 1.2 \end{bmatrix},$$

$$H^2 = \begin{bmatrix} 0.4 & -0.15 \\ -0.15 & 0.3 \end{bmatrix},$$

and

$$H^3 = \begin{bmatrix} 1.6 & -0.9 & 0.6 \\ -0.9 & 1.6 & -0.6 \\ 0.6 & -0.6 & 2.1 \end{bmatrix}.$$

We compute

$$H = H^1 \oplus H^2 \oplus H^3,$$

and

$$Q = H - A^T H A = \begin{bmatrix} 0.1480 & 0.0927 & -0.0230 & 0.0580 & 0.0502 & -0.0346 & -0.0380 \\ 0.0927 & 0.6784 & -0.0468 & -0.1015 & 0.0263 & -0.0094 & 0.1510 \\ -0.0230 & -0.0469 & 0.3068 & -0.0770 & -0.0695 & 0.0592 & -0.0565 \\ 0.0580 & -0.1016 & -0.0770 & 0.0920 & 0.1298 & -0.0679 & -0.0195 \\ 0.0503 & 0.0263 & -0.0695 & 0.1298 & 1.4500 & -0.7525 & 0.7323 \\ -0.0346 & -0.0094 & 0.0592 & -0.0679 & -0.7525 & 0.4290 & -0.4773 \\ -0.0380 & 0.1510 & -0.0565 & -0.0195 & 0.7323 & -0.4773 & 0.8400 \end{bmatrix}.$$

Since Q above is positive definite, all conditions of Theorem 7.3 are satisfied, and the equilibrium $x_e = 0$ of the 3–D system described by (7.51) using *saturation* arithmetic with coefficient matrix given in (7.57) is globally asymptotically stable. ∎

7.6 Concluding Remarks

In the present chapter we established several sufficient conditions for the global asymptotic stability of the equilibrium $x_e = 0$ of 2–D digital filters subject to overflow nonlinearities described by equation (7.1) (Theorem 7.1, Corollary 7.1, Proposition 7.1, and Theorem 7.2). The class of overflow nonlinearities which we considered herein include as special cases the usual types of overflow arithmetic employed in practice, including zeroing, two's complement, triangular, and saturation overflow characteristics. The stability results developed herein make use of a general class of positive definite and radially unbounded Lyapunov functions (Theorem 7.1, *i.e.*, Assumption (A-7.1)). Two special cases of these functions, including quadratic Lyapunov functions (Theorem 7.2) and l_p vector norms (Corollary 7.1 and Proposition 7.1), are considered. For quadratic forms, we presented necessary and sufficient conditions which enable us to construct the Lyapunov functions (Lemma 7.1). One of the results presented herein (Theorem 7.2) constitutes a generalization to one of the existing stability results (condition (7.3)) for 2–D digital filters described by (7.1).

Generalizations of the above results to m–D digital filters ($m > 2$) were also achieved (Theorem 7.3, Corollary 7.2, and Remark 7.10).

The results developed herein yield also conditions for the non-existence of limit cycles (under zero-input conditions) of 2–D and m–D fixed-point digital filters ($m > 2$) subject to overflow nonlinearities.

To demonstrate the applicability of the present results, we considered two specific examples (Examples 7.1 and 7.2).

PART III
ANALYSIS AND SYNTHESIS OF A CLASS OF NEURAL NETWORKS
WITH INTERCONNECTION CONSTRAINTS WITH APPLICATIONS TO
ASSOCIATIVE MEMORIES

CHAPTER 8

INTRODUCTION TO PART III

8.1   Models of Feedback Neural Networks

In [45], a feedback neural network model which is fully interconnected, known as the analog Hopfield model, is presented. Prior to Hopfield's work, this network had already been studied widely for over a decade under the name of *additive model* (see, e.g., [40], page 23). Hopfield's work may be viewed as one of the first significant applications of the additive model. In [45], Hopfield was able to provide a circuit realization for the additive model and was able to show that the trajectories of these networks would not oscillate, but would seek local minima of a certain type of energy function. The main assumption for this network is that the interconnection matrix must be symmetric and must have zero diagonal elements. Using this energy function as the basis for a design procedure, Hopfield neural networks have been utilized to convert analog signals to digital signals, to decompose additive signals, and to solve certain optimization problems (see [46], [112]).

The analog Hopfield neural network model is described by differential equations of the form

$$\dot{x} = -Ax + Ty + I, \tag{8.1}$$

where $x \in R^n$, $A = \text{diag}[a_1, \cdots, a_n]$ with $a_i > 0$, $T \in R^{n \times n}$, $I \in R^n$,

$$y = g(x) = [g_1(x_1), \cdots, g_n(x_n)]^T, \tag{8.2}$$

and $g_i \colon R \to (-1, 1)$ is called the activation function and is continuously differentiable, strictly monotonically increasing in x_i, and $x_i g(x_i) > 0$ for all $x_i \neq 0$.

In the literature it is frequently assumed that $T_{ij} = T_{ji}$ for all $i, j = 1, \cdots, n$ and that $T_{ii} = 0$ for all $i = 1, \cdots, n$. We will make these assumptions only when explicitly stated. In addition to (8.1), there are many other feedback neural network models which have been investigated in the literature. For example, neural networks described by

$$x(k+1) = \text{sign}(Tx(k) + I) \tag{8.3}$$

are considered in [44] and [83], where sign($\cdot$) is the signum function. Also, neural networks described by

$$\dot{x} = h(Tx + I) \tag{8.4}$$

are investigated in [59], where $h(\cdot)$ is defined in (3.1). Moreover, neural networks described by

$$x(k+1) = \text{sat}(Tx(k) + I) \tag{8.5}$$

are investigated in [87], where sat($\cdot$) is defined in (3.11). Furthermore, neural networks described by

$$x(k+1) = Tg(x(k)) + I \tag{8.6}$$

are considered in [75]. In the literature, neural network (8.3) is called the *discrete-time Hopfield model*, (8.4) and (8.5) are called *neural networks defined on hypercubes*, and (8.6) is called the *iterated map neural network*.

Some of the results concerning qualitative issues of neural networks described by (8.1), (8.3)–(8.6) are summarized below.

(1) Under the assumption that $T_{ij} = T_{ji}$ and $T_{ii} = 0$, Hopfield showed that for neural network (8.1), the function defined by

$$E(y) = -\frac{1}{2} \sum_{i=1}^{n} \sum_{j=1,j\neq i}^{n} T_{ij} y_i y_j + \sum_{i=1}^{n} a_i \int_0^{y_i} g_i^{-1}(\xi) d\xi - \sum_{i=1}^{n} I_i y_i \tag{8.7}$$

satisfies

$$\frac{dE}{dt} \leq 0 \quad \text{and} \quad \frac{dE}{dt} = 0 \rightarrow \frac{dy_i}{dt} = 0 \quad \text{for all } i.$$

This implies that the time evolution of the system is a motion in the state space which seeks out minima of the function E [45]. E defined in (8.7) is a Lyapunov function (or *energy function*) for the system. In [34] and [82], Michel and his coworkers conducted a systematic qualitative analysis of (8.1) using the techniques advanced in [86] and presented a synthesis procedure for (8.1) without the symmetry and zero-diagonal constraints on matrix T. (This synthesis procedure, as well as several other procedures will be summarized in the next section.)

(2) For neural network (8.3) with $I = 0$, it is shown in [44] that the energy function defined by

$$E(x) = -\frac{1}{2} \sum_{i=1}^{n} \sum_{j=1, j \neq i}^{n} T_{ij} x_i x_j$$

is monotonically decreasing when $T_{ij} = T_{ji}$ and $T_{ii} = 0$. This implies that the state of the system will continue to change until a local minimum of E is reached. Hopfield proposed a synthesis procedure for neural network (8.3) which is known as the outer product method. For neural network (8.3) with $I \not\equiv 0$ and without symmetry and zero-diagonal constraints, a systematic analysis and synthesis approach is developed by Michel *et al* in [83]. The analysis results in [83] are developed by adapting the techniques advanced in [86] and the synthesis procedure is referred to as the pseudo-inverse method with stability constraints.

(3) A qualitative analysis and synthesis of neural network (8.4), in which T is symmetric, are established in [59]. The analysis results presented in [59] enable us to locate all the equilibrium points and to ascertain the qualitative properties of these equilibria for a given neural network of the form (8.4). The synthesis approach developed in [59] for neural network (8.4) is referred to as the eigenstructure method. An energy function E is specified for this type of neural network. It is shown in [59] that there exists a one-to-one correspondence between the set of local minima of E and the set of asymptotically stable equilibrium points of a given neural network.

(4) Neural network (8.5) is the discrete-time counterpart of neural network (8.4). Qualitative analysis and synthesis results for (8.5) which are discrete-time counterparts of the results in [59] are established in [87]. Neural network (8.5) is also some times referred to as the brain-state-in-a-box (BSB) model [38], [39], [47]. The analyses given in [87] for this type of neural network with symmetric T are thorough and complete. Some analysis results for neural network (8.5) with non-symmetric T are presented in [39] and [47].

(5) The equilibrium points of (8.1) (with $A =$ the identity matrix) satisfy the condition

$$x = Ty + I = Tg(x) + I.$$

This relation motivates another neural network model–the iterated map neural networks described by equation (8.6). When T is symmetric, it is shown in [75] that the only attractors of (8.6) are fixed points (equilibrium points) and period-two limit cycles and that under certain conditions (8.6) has only fixed point attractors (*i.e.*, it is globally stable). An energy function is used to establish these results.

The g functions specified in (8.2) are sometimes called *sigmoidal functions*, which are approximations of nonlinearities in certain circuits. In the literature, one the most popular forms of function g_i is given by (see [82])

$$g_i(x_i) = \beta_i \tan^{-1}(\lambda_i x_i), \tag{8.8}$$

where $\beta_i > 0$ and $\lambda_i > 0$. However, in (8.1), it is sometimes desirable to use a function $g_i: R \to [-1, 1]$ which is not continuously differentiable. For example,

$$
\begin{aligned}
g_i(x_i) &= \text{sat}(x_i) \\
&= \tfrac{1}{2}(|x_i + 1| - |x_i - 1|)
\end{aligned}
\tag{8.9}
$$

is used in [22].

In the present monograph, we choose g_i as in equation (8.9), *i.e.*, $g(\cdot) = \text{sat}(\cdot)$, for neural network (8.1) (cf. the definition for function $\text{sat}(\cdot)$ on page 26). We will focus on applications to associative memories of neural networks described by equation (8.1) with $g(\cdot) = \text{sat}(\cdot)$. This type of network will be referred to as a neural network with (piecewise) linear saturation activation functions. We note that our results in Chapter 9 are adaptations of the results in [59] to the present neural network model, and that all of our analysis and synthesis results, to be established in Chapters 10 and 11, can be easily adapted to the neural networks described by equations (8.3)–(8.5) and to neural networks described by equation (8.6) with $g(\cdot) = \text{sat}(\cdot)$. We note that systems described by (8.4) and (8.5) are more general than those which we considered in Chapter 3 (cf. equations (3.1) and (3.11)). The reason for considering neural networks described by (8.1) is that such networks include cellular neural networks as special cases (see details in Section 10.3).

8.2   Design Methods for Associative Memories

Synthesis techniques for associative memories which have been devised for various neural network models include the outer product method [44], the projection learning rule [96], [97], the pseudo-inverse method with stability constraints [34], [82], [83], and the eigenstructure method [59], [87]. These methods have inherent advantages as well as limitations. Specifically:

The *Outer Product Method* [44] is capable of learning in the sense that one can add new patterns to be stored and is capable of forgetting in the sense that one can remove some patterns from the set of desired patterns, without having to recompute the *entire* interconnection matrix T. In addition, the networks designed by this method are globally stable (*i.e.*, all trajectories of the network will tend to some equilibrium point). On the other hand, a network obtained by this method may fail to store some of the desired patterns; its capacity of storing arbitrary patterns is limited to approximately $0.15n$ asymptotically stable equilibrium points (n denotes the order of the network), and its connection matrix is required to be symmetric. The latter restriction poses difficulties in implementations. Indeed, the inability to implement $T_{ij} = T_{ji}$ *precisely* may give rise to spurious states [5].

The *Projection Learning Rule* [96], [97] is capable of learning and forgetting in the same sense as the Outer Product Method. All solutions of a neural network designed by this algorithm will tend to some equilibrium point of the network. The projection learning rule guarantees to store specific prototype vectors as equilibrium points; however, these equilibria need not necessarily be asymptotically stable. The attractivity of patterns stored in neural networks (as asymptotically stable equilibria) designed by the projection learning rule appears to decrease sharply as the number of stored patterns approaches an order $0.5n$. As in the case of the Outer Product Method, the interconnection matrix of networks designed by the projection learning rule is required to be symmetric.

The *Pseudo-Inverse Method with Stability Constraints* [34], [82], [83] results in neural networks capable of storing up to n linearly independent equilibrium points. However, as in the case of the projection learning rule, the effectiveness of the attractivity of the stored equilibrium points is greatly reduced as the number of the desired vectors to be stored becomes of order $n/2$. This method guarantees that a network designed by this technique will always store a set of given vectors as asymptotically stable equilibrium points. The results in [121] make provisions of learning and forgetting for this method. Networks designed by this technique need not result in symmetric interconnecting structures; however, they are not necessarily globally stable.

The *Eigenstructure Method* [59], [87] is capable of learning and forgetting in the same sense as the above three methods [122]. All solutions of a neural network designed by this method will tend to some equilibrium point of the network. In a network designed by the eigenstructure method, *all* the desired patterns are guaranteed to be stored as *asymptotically stable* equilibrium points of the neural network. The number of patterns that can be stored as effective asymptotically stable equilibria in networks designed by the eigenstructure method can be significantly larger than by the above three methods. As in the preceding design methods, the interconnection matrix of neural networks designed by the eigenstructure method is required to be symmetric.

The above design methods usually result in neural networks which are *fully* interconnected. One of the noticeable deficiencies of these design methods, is their inability of generating networks with partial or sparse interconnections. In the third part of this monograph, we will remove this shortcoming from the eigenstructure method. Similar results can also be developed for the pseudo-inverse method (and projection learning rule); however, similar adaptations for the outer product method are not apparent at this time.

8.3   Analog VLSI Implementations and Problems

There are two general types of VLSI implementations for neural networks: digital and analog implementations. Compared to analog implementations, digital implementations are much easier to achieve through standard CMOS technology. Analog implementations, on the other hand, are more difficult to achieve, due to several problems. Networks implemented by analog VLSI technology can exhibit properties which digital implementations cannot achieve, including speed of the network evolution, less power consumption, reduced number of devices, and a close relation to the mathematical models used.

The symmetry constraints mentioned before, pose some difficulties in the implementations of feedback neural networks. However, the most severe difficulties encountered in the implementation process involve large numbers of interconnections and our inability of exact realizations of computed parameters. Current VLSI technology restricts the connectivity in neural network implementations to a level at which one cannot expect to achieve more than a few hundred neurons in an implemented neural network chip. Accordingly, for many problems, in which more neurons are required, fully connected neural networks are not feasible candidates. This has motivated another neural network structure, the cellular neural network structure, in which one considers only local interconnections which are restricted to a small neighborhood. However, for such networks, no successful design procedures for associative memories have been reported. This limits the applicability of cellular neural networks which are believed to be among the easiest to implement. Also, when implementing a designed neural network, one usually assumes that all the computed parameters are realized exactly, so that the network can exhibit a desired performance (such as having desired memory points). As can easily be imagined, almost every implementation process will result in some inaccuracies. When considering these inaccuracies, one has to have criteria that enable one to determine if the implemented neural network will perform as expected. The above observations suggest the following two important problems: (i) for a given set of desired memory patterns, we wish to design a neural network which does not require full interconnections; and (ii) for a given design problem, we seek criteria to determine bounds for the implementation inaccuracies of the computed parameters.

The objective of the present part of our research is to tackle the two important problems identified above. We will develop design procedures for neural networks with fewer interconnections and we will establish results which enable us to determine allowable upper bounds for the parameter inaccuracies in the implementation process. In particular, we will develop synthesis procedures which can be used to design neural networks with pre-specified partial or sparse interconnecting structure. We refer to such neural networks as *sparsely interconnected neural networks*. We will also establish analysis results so that for a given neural network, we can determine the permissible upper bounds for the perturbations on its parameters (e.g., A, T, and I in (8.1)). We will also develop a symmetric design procedure by means of which we can design a neural network with sparse and symmetric interconnection matrix. As a basis for the development of the above results, we will *first* establish results by means of which we can determine all the equilibrium points and their qualitative properties in a systematic manner for a given neural network; and we will *first* develop synthesis procedures for neural networks without any connectivity constraints.

8.4   Cellular Neural Networks

If we have $M \times N$ neurons in our neural network (8.1), we can always arrange these neurons in an $M \times N$ array. For example, when $M = 3$ and $N = 4$, we have the structure shown in Figure 8.1. Neural networks having the structure of Figure 8.1 are examples of (two-dimensional) *cellular neural networks*, introduced by Chua and Yang in 1988 [22]. The basic circuit unit of a cellular neural network, called a cell, contains linear and nonlinear circuit elements. Any cell in a cellular neural network is connected only to its neighboring cells, and thus, adjacent cells interact *directly* with each other. Cells that are not directly

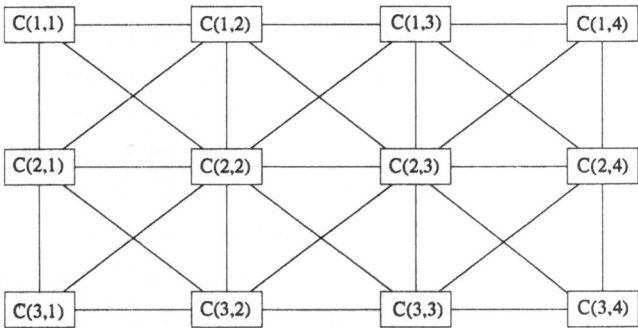

Figure 8.1: A two-dimensional cellular neural network structure

connected affect each other *indirectly* because of the propagation effects of the continuous-time dynamics in the network.

The cellular neural network of Figure 8.1 has a similar structure as a two-dimensional *cellular automaton*. Both have parallel signal processing capabilities and both have only local interconnections. However, a cellular neural network is usually required to seek stable equilibria, while a cellular automaton may exhibit many complex phenomena such as periodic and chaotic behaviors. The recent increasing interest in cellular neural networks is partly due to the fact that such networks are among the easiest to implement in (analog) VLSI. Several applications of cellular neural networks to image processing can be found in the literature. Applications of cellular neural networks to Chinese character recognition is reported in [23] where it is suggested that a particularly designed cellular neural network may have the capability of extracting some features from simple images such as Chinese characters. Several other applications of cellular neural networks to image processing, including hole-filler [76], connected components detector [77], shadow detector [78], and image thining [79] have also been reported. We will also consider applications of cellular neural networks to the Chinese character recognition problem (cf. Section 10.5 and [63]), however, from a different point of view. Specifically, we will utilize the capability of cellular neural networks to realize associative memories.

Although cellular neural networks have found several successful applications in image processing, their applications to associative memories are wanting. Since cellular neural networks are special cases of sparsely interconnected neural networks, our design procedures developed for sparsely interconnected neural networks can directly be applied to the design of cellular neural networks for associative memories.

In the applications of cellular neural networks to image processing [23], [76], [77], [78], [79], one has to choose carefully the dynamic rule specified by a *cloning template* (connection matrix). In the applications reported in [23], [76], [77], [78], [79], the cloning templates are chosen by experience. Since no general rules are known for choosing cloning templates, or equivalently, the connection matrix, for a given problem, great difficulties are encountered in the applications of cellular networks to problems of this type. Our applications of cellular neural networks to associative memories are more successful in the sense that for any given design problem, we have a systematic procedure to determine the connection matrix. To some extent, our work makes important original contributions to cellular neural networks both in theory and in practice.

CHAPTER 9

ANALYSIS AND SYNTHESIS OF A CLASS OF NEURAL NETWORKS
WITH PIECEWISE LINEAR SATURATION ACTIVATION FUNCTIONS

9.1 Introduction

Although artificial neural network research can be traced back three decades, it was during the last ten years that activities on neural networks have become a rapidly growing research area. A number of neural network models have been developed in the literature. These can roughly be divided into two categories: multi-layered and single-layered. Single-layered neural networks are usually fully connected, feedback type. This type of neural network is a good candidate for associative memories (content-addressable memories), optimization, image processing, and pattern recognition, where high speed parallel computations which are robust to malfunctions of individual devices are preferred.

The *qualitative analysis* of single-layered, fully connected neural networks has evoked a great deal of interest in recent years (see, e.g., [3], [13], [29], [38], [39], [41], [42], [47], [52], [57], [58], [59], [71], [75], [80], [82], [83], [84], [87], [101], [118]). These works are primarily concerned with the existence and the locations (in the state space) of equilibrium points, with the qualitative properties of the equilibria, and with the extent of the basins of attraction of asymptotically stable equilibria (memories). The stability analysis of a class of single layered, *partially* interconnected neural networks–*Cellular Neural Networks,* has also been of recent interest (see, for example, [18], [21], [22], [65], [66], [70], [72], [73], [100], and [123]).

In [20], [30], [34], [44], [57], [58], [59], [63], [65], [66], [70], [72], [73], [83], [84], [87], [95], [96], [97], [121], [122], several synthesis procedures, such as the *outer product method* [44], the *pseudo-inverse method* (also called the *projection learning rule*) [34], [83], [96], [97], [121], the *eigenstructure method* [57], [58], [59], [63], [65], [66], [70], [72], [73], [87], [122], and a few other methods [20], [30], [84], [95] are developed for different types of neural network models (for an overview of some of these procedures, see [81]). Among these synthesis techniques, the *eigenstructure method* appears to be especially effective. This method has successfully been applied to the synthesis of neural networks defined on hypercubes [59], [87] and the Hopfield model [57], [58], [59], [104], and it can also be implemented iteratively [122].

In the present chapter, we consider neural networks described by equations of the form

$$\begin{cases} \dot{x} = -Ax + T\text{sat}(x) + I \\ y = \text{sat}(x) \end{cases} \tag{9.1}$$

where $x \in R^n$ is the state vector, $\dot{x}$ denotes the derivative of x with respect to time t, $y \in D^n$ is the output vector (D^n is defined on page 19 and represents the closed unit hypercube),

$$A = \text{diag}[a_1, \cdots, a_n]$$

with $a_i > 0$ for $i = 1, \cdots, n$, $T = [T_{ij}] \in R^{n \times n}$ is the connection (or coefficient) matrix, $I = [I_1, \cdots, I_n]^T \in R^n$ is a bias vector, and

$$\text{sat}(x) = [\text{sat}(x_1), \cdots, \text{sat}(x_n)]^T$$

represents the activation function, where $\text{sat}(\cdot)$ is defined in Equation (3.11) on page 26. We assume that the initial states of (9.1) satisfy $|x_i(0)| \leq 1$ for $i = 1, \cdots, n$.

System (9.1) is a variant of the analog Hopfield model (8.1) with a piecewise linear activation function $\text{sat}(\cdot)$. In the analog Hopfield model [45], one requires that T be symmetric. We do not make this assumption for (9.1). In this chapter and in the subsequent two chapters, we concern ourselves primarily with the implementation of *associative memories* by means of artificial neural networks (modeled by (9.1)).

One way of implementing system (9.1) is to consider the block diagram given in Figure 9.1. This figure suggests a standard control system structure—a feedback control system with a constant input and with output saturation nonlinearities. Based on the diagram in Figure 9.1, one can consider many different circuit realizations. For example, the circuit realizations proposed in [22] (for the cellular neural network model), [45] (for the analog Hopfield model), and [59] (for the neural networks defined on hypercubes), are all applicable (after slight modifications) to the present neural network model described by (9.1).

In Section 9.3, we will employ the techniques developed in [59], [87], to establish results concerning the qualitative properties of neural networks described by (9.1). Using the

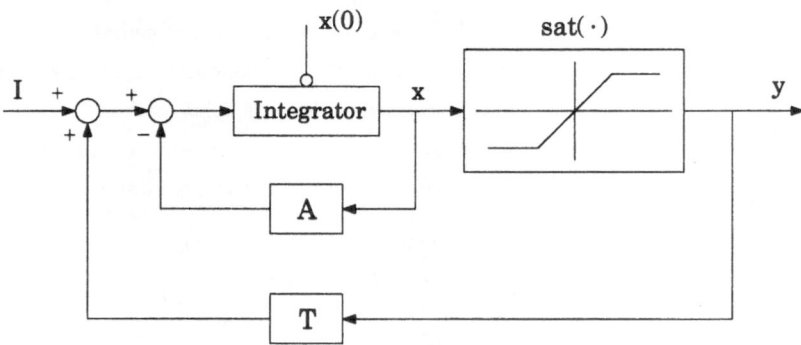

Figure 9.1: The block diagram of system (9.1)

results of Section 9.3, we will develop in Section 9.4 synthesis procedures for such systems. We consider a specific example in Section 9.5 and we conclude this chapter with several pertinent remarks in Section 9.6.

In our example (Section 9.5), we will conduct simulations of system (9.1), using the difference equations

$$
\begin{cases}
x_i((k+1)h) = [x_i(kh) + h\sum_{j=1}^{n} T_{ij}y_j(kh) + \frac{I_i}{a_i}(e^{a_i h} - 1)]e^{-a_i h} \\
y_i(kh) = \mathrm{sat}(x_i(kh))
\end{cases}
, \quad k = 0, 1, \cdots, \qquad (9.2)
$$

$i = 1, \cdots, n$, where h is the step size.

9.2 Notation

In the present section, we provide some notation which will be used in the sequel.

Let V and W be arbitrary sets. Then $V \cup W$, $V \cap W$, $V - W$, and $V \times W$ denote the union, intersection, difference, and Cartesian product of V and W, respectively. If V is a subset of W, we write $V \subset W$ and if x is an element of V, we write $x \in V$. If f is a function from V into W, we write $f: V \to W$ and we let $f(U) = \{f(x) \in W : x \in U\}$ for $U \subset V$, and $f^{-1}(y) = \{x \in V : f(x) = y\}$ for $y \in W$. Let ϕ denote the empty set. Let R denote the set of real numbers and let $R^+ = [0, +\infty)$. If $V_1, \cdots, V_n$ are n arbitrary sets, their Cartesian product is denoted by $\Pi_{i=1}^{n} V_i = V_1 \times \cdots \times V_n$. If in particular, $V = V_1 = \cdots = V_n$, we write $\Pi_{i=1}^{n} V_i = V^n$. Let R^n be real n-space. If $x \in R^n$, then $x^T = [x_1, \cdots, x_n]$ denotes the transpose of x. If $x \in R^n$ and $Y \subset R^n$, then $x \perp Y$ will mean that $x^T \cdot y = 0$ for all $y \in Y$. If $V \subset R^n$, then $\overline{V}, V^0$ and ∂V represent the closure, interior and boundary of V in R^n, respectively. Also, we let $B(\tilde{x}, r) = \{x \in R^n : \|\tilde{x} - x\| < r\}$ for $\tilde{x} \in R^n$ and $r > 0$. Let $B^n = \{x \in R^n : x_i = 1$ or $-1, i = 1, \cdots, n\}$ and $D^n = \{x \in R^n : -1 \le x_i \le 1, i = 1, \cdots, n\}$.

If $A = [A_{ij}]$ is an arbitrary matrix, then A^T denotes the transpose of A. If A is a symmetric matrix, by $A > 0$ we mean that A is positive definite and by $A \ge 0$ we mean that A is positive semidefinite. If A is a square matrix, we use $\lambda(A)$ to denote eigenvalues of A. Let $P(n)$ denote the set of all permutations on $\{1, \cdots, n\}$. If $\{x_1, \cdots, x_m\} \subset R^n$, then $\text{Span}(x_1, \cdots, x_m)$ denotes the linear subspace of R^n generated by $x_1, \cdots, x_m$ and $\text{Aspan}(x_1, \cdots, x_m)$ denotes the affine subspace of R^n generated by $x_1, \cdots, x_m$. If $x_0 \in R^n$ and L is a linear subspace of R^n, then $L + x_0$ denotes the affine subspace of R^n produced by shifting L by x_0, that is, $L + x_0 = \{y \in R^n : y = x + x_0, x \in L\}$. In particular, $\text{Aspan}(x_1, \cdots, x_m) = \text{Span}(x_1 - x_m, \cdots, x_{m-1} - x_m) + x_m$.

9.3 Analysis Results

In this section, we establish results which characterize the qualitative behavior of system (9.1) by utilizing techniques similar to those developed in [59], [87]. We first introduce the following notation (which is slightly different from that used in Section 3.3).

For each integer m, $0 \le m \le n$, let

$$\Lambda_m = \{\xi = [\xi_1, \cdots, \xi_n]^T \in \Lambda : \xi_{\sigma(i)} = 0, 1 \le i \le m, \text{ and } \xi_{\sigma(i)} = \pm 1, m < i \le n,$$

$$\text{for some } \sigma \in P(n)\}$$

where

$$\Lambda = \{\xi = [\xi_1, \cdots, \xi_n]^T : \xi_i = \pm 1 \text{ or } 0, 1 \le i \le n\}$$

and $P(n)$ denotes the set of all permutations on $\{1, \cdots, n\}$. (Recall that there are $n!$ elements in $P(n)$.) For each $\xi \in \Lambda$, let

$$C(\xi) = \{x = [x_1, \cdots, x_n]^T \in R^n : |x_i| < 1 \text{ if } \xi_i = 0, \ x_i \ge 1 \text{ if } \xi_i = 1,$$

$$\text{and } x_i \le -1 \text{ if } \xi_i = -1\}.$$

From the notation given above, we have the following result.

Lemma 9.1

1) $\Lambda = \overset{n}{\underset{m=0}{\bigcup}} \Lambda_m$.

2) $\Lambda_0 = B^n$ and

$$C(\xi) = \{x \in R^n : |x_i| \geq 1,\ x_i \xi_i > 0,\ i = 1, \cdots, n\}$$

for any $\xi \in \Lambda_0$.

3) $\Lambda_n = \{0\}$ and

$$C(0) = (D^n)^0 = \{x \in R^n : -1 < x_i < 1,\ i = 1, \cdots, n\}.$$

4) $R^n = \overset{n}{\underset{m=0}{\bigcup}} \{C(\xi),\ \xi \in \Lambda_m\}$.

5) For any $\xi,\ \eta \in \Lambda,\ \xi \neq \eta,\ C(\xi) \cap C(\eta) = \phi$. ∎

Suppose that $\xi \in \Lambda_m$ and $\sigma \in P(n)$ such that

$$\xi_{\sigma(i)} = 0,\ 1 \leq i \leq m \ \text{ and } \ \xi_{\sigma(i)} = \pm 1,\ m < i \leq n. \tag{9.3}$$

We denote

$$A_I = \text{diag}[a_{\sigma(1)}, \cdots, a_{\sigma(m)}],$$

$$A_{II} = \text{diag}[a_{\sigma(m+1)}, \cdots, a_{\sigma(n)}],$$

$$T_{I,I} = [T_{\sigma(i)\sigma(j)}]_{1 \leq i,j \leq m},$$

$$T_{I,II} = [T_{\sigma(i)\sigma(j)}]_{1 \leq i \leq m, m < j \leq n},$$

$$T_{II,I} = [T_{\sigma(i)\sigma(j)}]_{m < i \leq n, 1 \leq j \leq m},$$

$$T_{II,II} = [T_{\sigma(i)\sigma(j)}]_{m < i,j \leq n},$$

$$I_I = [I_{\sigma(1)}, \cdots, I_{\sigma(m)}]^T,$$

$$I_{II} = [I_{\sigma(m+1)}, \cdots, I_{\sigma(n)}]^T,$$

$$\xi_I = [\xi_{\sigma(1)}, \cdots, \xi_{\sigma(m)}]^T,$$

and

$$\xi_{II} = [\xi_{\sigma(m+1)}, \cdots, \xi_{\sigma(n)}]^T.$$

Remark 9.1 In the notation introduced above, $C(\xi)$, for each $\xi \in \Lambda_m$, $0 \le m \le n$, represents different disjoint regions in R^n, while the notation $C(\xi)$ used in [59], [87] (also see Section 3.3) is defined for disjoint regions on the closed hypercube D^n. ∎

Remark 9.2 For a given $\xi \in \Lambda_m$, there may exist different elements in P(n) for which (9.3) is true. For these different permutations, the notation given above will be the same up to different orders in the components. Thus, the subsequent analysis and conclusions will be identical for any of the permutations used. ∎

Remark 9.3 If $m = n$, we have $A_I = A$, $T_{I,I} = T$, $I_I = I$, $\xi_I = \xi$ and the A_{II}, $T_{I,II}$, $T_{II,I}$, $T_{II,II}$, I_{II}, ξ_{II} do not exist. If $m = 0$, we have $A_{II} = A$, $T_{II,II} = T$, $I_{II} = I$, $\xi_{II} = \xi$ and the A_I, $T_{I,I}$, $T_{I,II}$, $T_{II,I}$, I_I, ξ_I do not exist. ∎

Consider $\xi \in \Lambda_m$, $0 < m < n$, with $\sigma \in$ P(n) such that $\xi_{\sigma(i)} = 0$, $1 \le i \le m$, and $\xi_{\sigma(i)} = \pm 1$, $m < i \le n$. We can rewrite the first equation of system (9.1) as

$$\begin{cases} \dot{x}_I = -A_I x_I + T_{I,I} x_I + T_{I,II} \xi_{II} + I_I \\ \dot{x}_{II} = -A_{II} x_{II} + T_{II,I} x_I + T_{II,II} \xi_{II} + I_{II} \end{cases} \tag{9.4}$$

where

$$\xi_{II} = [\xi_{\sigma(m+1)}, \cdots, \xi_{\sigma(n)}]^T,$$

$$x_I = [x_{\sigma(1)}, \cdots, x_{\sigma(m)}]^T \text{ with } -1 < x_{\sigma(i)} < 1 \text{ for } 1 \le i \le m,$$

and

$$x_{II} = [x_{\sigma(m+1)}, \cdots, x_{\sigma(n)}]^T \text{ with } \xi_{\sigma(i)} x_{\sigma(i)} \ge 1 \text{ for } m < i \le n.$$

Equation (9.4) is said to be an *equivalent linear representation of system (9.1) over the region* $C(\xi)$.

When $m = n$, $\Lambda_n = \{0\}$. In this case, for $x \in C(0) = (D^n)^0$, system (9.1) becomes

$$\dot{x} = (T - A)x + I. \tag{9.5}$$

When $m = 0$, $\Lambda_0 = B^n$. In this case, for $\xi \in \Lambda_0$, $x \in C(\xi)$, system (9.1) can be expressed as

$$\dot{x} = -Ax + T\xi + I. \tag{9.6}$$

We will have occasion to make use of the following hypotheses for system (9.1).

Assumption (A–9.1) For any m, $0 < m \leq n$, and for any $\xi \in \Lambda_m$, the $m \times m$ matrix

$$T_{I,I} - A_I = [T_{\sigma(i)\sigma(j)}]_{1 \leq i,j \leq m} - \text{diag}[a_{\sigma(1)}, \cdots, a_{\sigma(m)}]$$

is non-singular, where $\sigma \in P(n)$ so that $\xi_{\sigma(i)} = 0$, $1 \leq i \leq m$ and $\xi_{\sigma(i)} = \pm 1$, $m < i \leq n$. ∎

For system (9.1) satisfying Assumption (A–9.1) we will employ the following notation.

1) For $\xi = 0 \in \Lambda_n$ $(m = n)$, let

$$x_\xi = (A - T)^{-1}I. \tag{9.7}$$

2) For $\xi \in \Lambda_m$, $0 < m < n$, with A_I, A_{II}, $T_{I,I}, \cdots, T_{II,II}$, I_I, I_{II} defined above, let

$$x_\xi = [x_{\xi 1}, \cdots, x_{\xi n}]^T \in R^n \tag{9.8}$$

where

$$x_{\xi I} = [x_{\xi\sigma(1)}, \cdots, x_{\xi\sigma(m)}]^T = (A_I - T_{I,I})^{-1}(T_{I,II}\xi_{II} + I_I),$$

and

$$x_{\xi II} = [x_{\xi\sigma(m+1)}, \cdots, x_{\xi\sigma(n)}]^T = A_{II}^{-1}(T_{II,I}x_{\xi I} + T_{II,II}\xi_{II} + I_{II}).$$

3) For $\xi \in \Lambda_0 = B^n$ $(m = 0)$, let

$$x_\xi = A^{-1}(T\xi + I). \tag{9.9}$$

4) For the x_ξ defined above, let

$$y_\xi = \text{sat}(x_\xi).$$

Assumption (A–9.2) With the notation given above, assume that for any $\xi \in \Lambda_m$, $0 \leq m \leq n$, $x_\xi \notin \partial(C(\xi))$. ∎

The following result enables us to locate in a systematic manner all equilibria for system (9.1) and to ascertain the stability properties of these equilibria. Furthermore, this result will serve as the theoretical basis of the synthesis procedures which we will present in the sequel.

Theorem 9.1 Suppose that (9.1) satisfies Assumptions (A–9.1) and (A–9.2). For any $m, 0 \leq m \leq n$, and for any $\xi \in \Lambda_m$, we have:

Case I: $m = n$, $\xi = 0 \in \Lambda_n$. (Note that in this case $C(\xi) = (D^n)^0$.)

1) If $x_\xi \notin (D^n)^0$, there is no equilibrium point of system (9.1) in $(D^n)^0$.

2) If $x_\xi \in (D^n)^0$, x_ξ is the unique equilibrium point of system (9.1) in $(D^n)^0$. In particular,

(i) if $T - A$ has one or more eigenvalues with non-negative real parts, x_ξ is unstable, and

(ii) if all eigenvalues of $T - A$ have negative real parts, x_ξ is asymptotically stable.

Case II: $0 < m < n$, $\xi \in \Lambda_m$.

1) If $x_\xi \notin C(\xi)$, there is no equilibrium point of system (9.1) in $C(\xi)$.

2) If $x_\xi \in C(\xi)$, x_ξ is the unique equilibrium point of system (9.1) in $C(\xi)$. In particular,

(i) if $T_{I,I} - A_I$ has one or more eigenvalues with non-negative real parts, x_ξ is unstable, and

(ii) if all eigenvalues of $T_{I,I} - A_I$ have negative real parts, x_ξ is asymptotically stable.

Case III: $m = 0$, $\xi \in \Lambda_0 = B^n$.

1) If $x_\xi \notin C(\xi)$, there is no equilibrium point of system (9.1) in $C(\xi)$.

2) If $x_\xi \in C(\xi)$, x_ξ is an asymptotically stable equilibrium point of system (9.1).

Proof: For each $\xi \in \Lambda_m$, $0 \leq m \leq n$, consider (9.4)–(9.6). Using similar arguments as in [59], the conclusions of this theorem follow directly from the theory of linear differential equations (cf., for example, [88]). ∎

If x_ξ is an asymptotically stable equilibrium point of system (9.1), $y_\xi = \mathrm{sat}(x_\xi)$ is said to be a *memory vector* of system (9.1). A memory vector y_ξ is said to be *reachable* if there exists a neighborhood V of y_ξ such that for any $x(0) \in V \cap D^n \neq \phi$, the output vector $y(t)$ of system (9.1) tends to y_ξ asymptotically as $t \to \infty$. Using the results given in Theorem 9.1, it can easily be shown that a memory vector $y_\xi \in (D^n)^0$ or $y_\xi \in B^n$ is always reachable. When a memory vector $y_\xi \in \partial(D^n) - B^n$, y_ξ is reachable if and only if there exists a neighborhood U of y_ξ, such that the set $U \cap D^n$ has a non-empty intersection with the domain of attraction of the corresponding asymptotically stable equilibrium point x_ξ. In our synthesis procedures, the objective is to store patterns in B^n. If we can guarantee that a desired set of bipolar patterns is stored as a set of memory vectors, then such vectors will always be reachable. Therefore, in the sequel, we will drop the modifier "reachable" when the context is clear.

Remark 9.4 With T symmetric, the function $E: D^n \to R$ defined by

$$E(y) = -\frac{1}{2}y^T T y + \frac{1}{2}y^T A y - y^T I \tag{9.10}$$

can be shown to be monotonically decreasing in time t along the solutions of (9.1). To see this, we follow the same procedure as in [59] by defining a *local solution* for system (9.1); or, we can follow the procedure in [22], by defining the derivatives dy_i/dx_i at the breaking points $|x_i| = 1$ to be zero. This guarantees that system (9.1) will neither oscillate nor become chaotic. When T is nonsymmetric, the function E defined in (9.10) is not necessarily monotonically decreasing, and oscillatory solutions for (9.1) may exist. ∎

Remark 9.5 It is possible to generalize Theorem 9.1 to a result which does not require Assumption (A-9.2). However, for such a case, the conclusions of the theorem will be less straightforward. ∎

Several important conclusions can be drawn from Theorem 9.1 which are given in the following.

Corollary 9.1 Suppose that in system (9.1) T is symmetric and that the coefficients of system (9.1) satisfy the conditions that

$$T_{ii} > a_i \text{ for } i = 1, \cdots, n. \tag{9.11}$$

Then, every asymptotically stable equilibrium point $x_e = [x_{e1}, \cdots, x_{en}]^T$ of system (9.1) satisfies the condition that

$$|x_{ei}| > 1, \text{ for } i = 1, \cdots, n. \tag{9.12}$$

If (9.11) is satisfied, $T_{I,I} - A_I$ and $T - A$ will have one or more non-negative eigenvalues, since $T_{I,I} - A_I$ and $T - A$ are symmetric matrices with positive diagonal elements. Thus, stability conditions of an equilibrium in Cases I and II of Theorem 9.1 can never be satisfied. The only stability conditions for (9.1) which may be satisfied are those that apply to Case III. It is clear that every equilibrium of (9.1) which satisfies the conditions in Case III of Theorem 9.1 has the property given in (9.12). ∎

We point out that the same result as Corollary 9.1 has been proved for cellular neural networks (with *symmetric* interconnections) in [22], using a different approach.

Corollary 9.2 Every asymptotically stable equilibrium point $x_e = [x_{e1}, \cdots, x_{en}]^T$ of (9.1) satisfies condition (9.12) if

$$T_{ii} > a_i + \sum_{j=1, j\neq i}^{n} |T_{ij}|, \text{ for } i = 1, \cdots, n. \tag{9.13}$$

If (9.13) is satisfied, $T_{I,I} - A_I$ and $T - A$ become matrices with positive diagonal elements and satisfy a diagonal dominance condition. By Gerŝgorin's Theorem [74], all eigenvalues of $T_{I,I} - A_I$ and $T - A$ are contained in the union of the n disks of the complex plane centered at $T_{ii} - a_i$ with radius

$$\sum_{j=1, j\neq i}^{n} |T_{ij}|.$$

This in turn implies that $Re\lambda(T_{I,I} - A_I)$ and $Re\lambda(T - A)$ are positive. From Theorem 9.1, we see that (9.12) is satisfied. ∎

Corollary 9.3 Suppose that β is an asymptotically stable equilibrium point and $\alpha = \text{sat}(\beta)$ is a memory vector of system (9.1) with parameters $\{A, T, I\}$. Then, α and β will also be a pair of memory vector and asymptotically stable equilibrium point of system (9.1) with parameters $\{kA, kT, kI\}$ for every real number $k > 0$.

The proof can easily be established by considering Equations (9.7)–(9.9) and Theorem 9.1. ∎

Remark 9.6 The significance of Corollary 9.3 is that for a given neural network (9.1), we can increase its speed of evolution by multiplying A, T, and I by a constant $k > 1$, without changing any of its asymptotically stable equilibrium points and any of its memory vectors. Since the speed of evolution of (9.1) depends on the eigenvalues of $T - A$ and $T_{I,I} - A_I$, it is also clear that the larger the k is, the faster the evolution will be. ∎

9.4 Synthesis Procedures

In this section, we present synthesis procedures for system (9.1). As pointed out earlier, Theorem 9.1 will serve as the basis for the synthesis procedures developed in the present chapter. In particular, we point to the following important fact, which is a consequence of Case III in Theorem 9.1.

Corollary 9.4 Suppose $\alpha \in B^n$. If

$$\beta = A^{-1}(T\alpha + I) \in (C(\alpha))^0,$$

then β is an asymptotically stable equilibrium point of (9.1). ∎

We are now in a position to address the following synthesis problem.

Synthesis Problem: Given m vectors in B^n (desired memory patterns), say $\alpha^1, \cdots, \alpha^m$, how can we properly choose $\{A, T, I\}$ so that the resulting synthesized system (9.1) has the properties enumerated below?

1) $\alpha^1, \cdots, \alpha^m$ are memory vectors of system (9.1).

2) The system has no oscillatory solutions.

3) The total number of spurious memory vectors (*i.e.*, memory vectors of (9.1) contained in $D^n - \{\alpha^1, \cdots, \alpha^m\}$) is as small as possible.

Remark 9.7 In practice, it is usually required that memory patterns be bipolar, *i.e.*, the memory patterns are in B^n. We will not consider the case where desired memory patterns are not in B^n. ∎

The preceding results allow us to approach the above synthesis problem in the following manner.

Synthesis Strategy: Given m vectors $\alpha^1, \cdots, \alpha^m$ in B^n, find A, T, and I such that

1) $A = \text{diag}[a_1, \cdots, a_n]$ with $a_i > 0$.

2) T is symmetric and has repeated eigenvalues equal to $\mu > 0$ and $-\tau < 0$.

3) $A\beta^i = T\alpha^i + I$ and $A\beta^i = \mu\alpha^i$, where $\beta^i \in (C(\alpha^i))^0$ and μ is the positive eigenvalue of T. ∎

In the following, we give the rationale for the above strategy.

1) From Remark 9.4, we see that when T is symmetric, system (9.1) will have no oscillatory solutions.

2) By Corollary 9.4,

$$A^{-1}(T\alpha^i + I) = \beta^i \in (C(\alpha^i))^0,$$

implies that β^i is an asymptotically stable equilibrium point of the synthesized system, and thus α^i is a memory vector.

3) For $\beta^i \in (C(\alpha^i))^0$ and $A\beta^i = \mu\alpha^i$, $i = 1, \cdots, m$, T and I are determined by the relations

$$A\beta^i = \mu\alpha^i = T\alpha^i + I, \quad i = 1, \cdots, m. \tag{9.14}$$

Solutions of (9.14) for T and I will always exist. To see this, we let

$$Y = [\alpha^1 - \alpha^m, \cdots, \alpha^{m-1} - \alpha^m].$$

We need to solve T from

$$TY = \mu Y, \tag{9.15}$$

and set

$$I = \mu\alpha^m - T\alpha^m.$$

Solutions of (9.15) for T will always exist. We will solve for T by using the singular value decomposition method.

We next present a solution to the Synthesis Problem based on the above observations.

Synthesis Procedure 9.1 Suppose we are given m vectors $\alpha^1, \cdots, \alpha^m$ in B^n, which are to be stored as memory vectors for (9.1). We proceed as follows:

1) Choose vectors $\beta^i \in (C(\alpha^i))^0$ for $i = 1, \cdots, m$, and a diagonal matrix A with positive diagonal elements, such that $A\beta^i = \mu\alpha^i$, where $\mu > 0$, i.e., choose $\beta^i = [\beta^i_1, \cdots, \beta^i_n]^T$ with $\beta^i_j\alpha^i_j > 1$ for $i = 1, \cdots, m$ and $j = 1, \cdots, n$, $A = \text{diag}[a_1, \cdots, a_n]$ with $a_j > 0$ for $j = 1, \cdots, n$, and

$$\mu > \max_{1 \le i \le n}\{a_i\}$$

such that $a_j\beta^i_j = \mu\alpha^i_j$.

2) Compute the $n \times (m-1)$ matrix:

$$Y = [y^1, \cdots, y^{m-1}] = [\alpha^1 - \alpha^m, \cdots, \alpha^{m-1} - \alpha^m]. \tag{9.16}$$

3) Perform a singular value decomposition of $Y = U\Sigma V^T$, where U and V are unitary matrices and Σ is a diagonal matrix with the singular values of Y on its diagonal. (This

can be accomplished by standard computer routines.) Let

$$U = [u^1, \cdots, u^n],$$

and

$$p = \text{dimension of Span}(y^1, \cdots, y^{m-1}).$$

From the properties of singular value decomposition, we know that $p = \text{rank}(Y)$, $\{u^1, \cdots, u^p\}$ is an orthonormal basis of $\text{Span}(y^1, \cdots, y^{m-1})$, and $\{u^1, \cdots, u^n\}$ is an orthonormal basis of R^n.

4) Compute

$$T^+ = [T^+_{ij}] = \sum_{i=1}^{p} u^i(u^i)^T, \text{ and } T^- = [T^-_{ij}] = \sum_{i=p+1}^{n} u^i(u^i)^T.$$

5) Choose a positive value for the parameter τ and compute

$$T_\tau = \mu T^+ - \tau T^- \text{ and } I_\tau = \mu \alpha^m - T_\tau \alpha^m. \tag{9.17}$$

Then, $\alpha^1, \cdots, \alpha^m$ will be stored as memory vectors in the following system

$$\begin{cases} \dot{x} = -Ax + T_\tau \text{sat}(x) + I_\tau \\ y = \text{sat}(x) \end{cases} \tag{9.18}$$

The states β^i corresponding to α^i, $i = 1, \cdots, m$, will be asymptotically stable equilibrium points of system (9.18). ∎

Remark 9.8 If we wish that in the synthesized system (9.18), the constant vector $I_\tau = 0$, we can modify Synthesis Procedure 9.1 as follows:

a) In step 2, take

$$Y = [\alpha^1, \cdots, \alpha^m]. \tag{9.19}$$

b) In step 5, take $I_\tau = 0$.

Then all conclusions will remain unchanged. In particular, each $-\beta^i$ and $-\alpha^i$, $i = 1, \cdots, m$, will also be asymptotically stable equilibrium points and memory vectors, respectively, of the synthesized system (9.18). ∎

Remark 9.9 The synthesis procedure presented above is an adaptation of the *eigenstructure method* developed in [59], [87] to the neural network model described by (9.1). Using the eigenstructure method, one can guarantee that *any* set of given memory patterns in B^n be stored as memory vectors. However, it is usually required that $p < n$, where $p = \text{rank}(Y)$, Y is defined in (9.16) or (9.19), and n is the order of the system. This follows, since if $p = n$, T_r in (9.17) will become a diagonal matrix with all diagonal elements equal to μ, and I_r in (9.17) becomes a zero vector, in which case all vectors in B^n (all corners of the hypercube D^n) will be stored as memory vectors. In such a case, the network is actually useless. Simulation results show that when p is very close to $n - 1$, there will be many spurious memory locations in the synthesized system. Experimental studies which compare the eigenstructure method (implemented on various types of artificial neural networks) with other methods have been conducted in several previous works (see, e.g., [57], [58], [59], [81], [96], [97], [122]). These works indicate that the capacity of neural network paradigms which make use of the eigenstructure method compare rather well with other paradigms (which make use, e.g., of the outer product method, pseudo-inverse techniques, and the like). ∎

Remark 9.10 Following the same procedure as in [59], [87], we can prove that all vectors in $L_\alpha \cap B^n$, where $L_\alpha = \text{Aspan}(\alpha^1, \cdots, \alpha^m)$, including $\alpha^1, \cdots, \alpha^m$, will be stored as memory vectors in system (9.18). ∎

9.5 An Example

To demonstrate the applicability of our present results, we consider the following example.

Example We use the synthesis procedure developed in this chapter to synthesize a neural network (9.1) with $n = 10$. Given are $m = 5$ vectors specified by (cf. [59], Example 6.1)

$$\alpha^1 = [-1,\ 1,\ -1,\ 1,\ 1,\ 1,\ -1,\ 1,\ 1,\ 1]^T,$$

$$\alpha^2 = [1,\ 1,\ -1,\ -1,\ 1,\ -1,\ 1,\ -1,\ 1,\ 1]^T,$$

$$\alpha^3 = [-1,\ 1,\ 1,\ 1,\ -1,\ -1,\ 1,\ -1,\ 1,\ -1]^T,$$

$$\alpha^4 = [1,\ 1,\ -1,\ 1,\ -1,\ 1,\ -1,\ 1,\ 1,\ 1]^T,$$

and

$$\alpha^5 = [1, -1, -1, -1, 1, 1, 1, -1, -1, -1]^T.$$

It is desired that these vectors be stored as memory vectors of system (9.1).

Using the Synthesis Procedure 9.1, we choose A as the 10×10 identity matrix, and determine

$$T = \begin{bmatrix}
-1.4161e+00 & -7.0073e-01 & -2.3650e+00 & -2.8029e+00 & -3.4161e+00 \\
-7.0073e-01 & -6.7591e+00 & 4.3796e-01 & 9.6350e-01 & -7.0073e-01 \\
-2.3650e+00 & 4.3796e-01 & -7.0219e+00 & 1.7518e+00 & -2.3650e+00 \\
-2.8029e+00 & 9.6350e-01 & 1.7518e+00 & -6.1460e+00 & -2.8029e+00 \\
-3.4161e+00 & -7.0073e-01 & -2.3650e+00 & -2.8029e+00 & -1.4161e+00 \\
2.6277e-01 & -2.7153e+00 & -1.6642e+00 & 1.1387e+00 & 2.6277e-01 \\
4.3796e-01 & -5.2555e-01 & 1.2263e+00 & -2.1022e+00 & 4.3796e-01 \\
-4.3796e-01 & 5.2555e-01 & -1.2263e+00 & 2.1022e+00 & -4.3796e-01 \\
-7.0073e-01 & 3.2409e+00 & 4.3796e-01 & 9.6350e-01 & -7.0073e-01 \\
1.6642e+00 & 2.8029e+00 & -2.5401e+00 & -7.8832e-01 & 1.6642e+00
\end{bmatrix}$$

$$\begin{bmatrix}
2.6277e-01 & 4.3796e-01 & -4.3796e-01 & -7.0073e-01 & 1.6642e+00 \\
-2.7153e+00 & -5.2555e-01 & 5.2555e-01 & 3.2409e+00 & 2.8029e+00 \\
-1.6642e+00 & 1.2263e+00 & -1.2263e+00 & 4.3796e-01 & -2.5401e+00 \\
1.1387e+00 & -2.1022e+00 & 2.1022e+00 & 9.6350e-01 & -7.8832e-01 \\
2.6277e-01 & 4.3796e-01 & -4.3796e-01 & -7.0073e-01 & 1.6642e+00 \\
-4.4818e+00 & -2.8029e+00 & 2.8029e+00 & -2.7153e+00 & -1.0511e+00 \\
-2.8029e+00 & -6.6715e+00 & -3.3285e+00 & -5.2555e-01 & -1.7518e+00 \\
2.8029e+00 & -3.3285e+00 & -6.6715e+00 & 5.2555e-01 & 1.7518e+00 \\
-2.7153e+00 & -5.2555e-01 & 5.2555e-01 & -6.7591e+00 & 2.8029e+00 \\
-1.0511e+00 & -1.7518e+00 & 1.7518e+00 & 2.8029e+00 & -4.6569e+00
\end{bmatrix},$$

and

$$I = [0.7883, 3.8540, -4.9927, 3.4161, 0.7883, 4.5547, 3.5912, -3.5912,$$

$$3.8540, -3.1533]^T.$$

(In the above computations, we chose $\mu = 2$ in step 1.)

Using Theorem 9.1, we verified that $\alpha^1, \cdots, \alpha^5$ are memory vectors of the synthesized system (9.1) with $\{A, T, I\}$ given above. Simulation results, using (9.2), also verified that $\alpha^1, \cdots, \alpha^5$ are reachable memory vectors of (9.1).

By Theorem 9.1, we determined that system (9.1) has no additional memory vectors in B^n. System (9.1) synthesized above has 8 memory vectors in $D^n - B^n$, given by

$$\alpha^5 = [-1,\ 1,\ 0,\ 0,\ 1,\ -1,\ 1,\ -1,\ 1,\ 0]^T,$$

$$\alpha^6 = [1,\ 1,\ 0,\ 0,\ -1,\ -1,\ 1,\ -1,\ 1,\ 0]^T,$$

$$\alpha^7 = [-1,\ 0,\ 0,\ 0,\ 1,\ 0,\ 1,\ -1,\ 0,\ -1]^T,$$

$$\alpha^8 = [1,\ 0,\ 0,\ 0,\ -1,\ 0,\ 1,\ -1,\ 0,\ -1]^T,$$

$$\alpha^9 = [0,\ 0,\ 0,\ 1,\ -1,\ 1,\ 0,\ 0,\ 0,\ -1]^T,$$

$$\alpha^{10} = [-1,\ -0.4563,\ -0.3634,\ 0.2612,\ 1,\ 1,\ 0.4663,\ -0.4663,\ -0.4563,\ -1]^T,$$

$$\alpha^{11} = [1,\ -0.4563,\ -0.3634,\ 0.2612,\ -1,\ 1,\ 0.4663,\ -0.4663,\ -0.4563,\ -1]^T,$$

and

$$\alpha^{12} = [-1,\ 0,\ 0,\ 1,\ 0,\ 1,\ 0,\ 0,\ 0,\ -1]^T.$$

Their corresponding asymptotically stable states are given by

$$\beta^5 = [-2,\ 2,\ 0,\ 0,\ 2,\ -2,\ 2,\ -2,\ 2,\ 0]^T,$$

$$\beta^6 = [2,\ 2,\ 0,\ 0,\ -2,\ -2,\ 2,\ -2,\ 2,\ 0]^T,$$

$$\beta^7 = [-2,\ 0,\ 0,\ 0,\ 2,\ 0,\ 2,\ -2,\ 0,\ -2]^T,$$

$$\beta^8 = [2,\ 0,\ 0,\ 0,\ -2,\ 0,\ 2,\ -2,\ 0,\ -2]^T,$$

$$\beta^9 = [0,\ 0,\ 0,\ 2,\ -2,\ 2,\ 0,\ 0,\ 0,\ -2]^T,$$

$$\beta^{10} = [-1.4380,\ -0.4563,\ -0.3634,\ 0.2612,\ 2.5620,\ 1.8908,\ 0.4663,$$
$$-0.4663 - 0.4563 - 3.0219]^T,$$

$$\beta^{11} = [2.5620,\ -0.4563,\ -0.3634,\ 0.2612,\ -1.4380,\ 1.8908,\ 0.4663,$$
$$-0.4663,\ -0.4563,\ -3.0219]^T,$$

and

$$\beta^{12} = [-2,\ 0,\ 0,\ 2,\ 0,\ 2,\ 0,\ 0,\ 0,\ -2]^T.$$

respectively. System (9.1) has also 69 unstable equilibrium points.

Digital simulations of (9.1) with A, T, and I determined above using equation (9.2), confirm all the above conclusions. ∎

9.6   Concluding Remarks

In the present chapter we considered a class of artificial neural networks which have the basic structure of the analog Hopfield neural networks [45] and which uses the (piecewise linear) saturation function to model the neurons (see Section 9.1 and system (9.1)). This model is closely related to the neural networks defined on hypercubes considered in [59].

For system (9.1), we first conducted a qualitative analysis which enables us to locate all of the equilibria and determine their stability properties (see Section 9.3 and Theorem 9.1). We established our results using a technique which is an adaptation of the technique advanced in [59] and [87].

Next, we developed for system (9.1) a synthesis procedure for associative memories which *guarantees* to store desired patterns in B^n as memories (see Section 9.4). The rationale for this procedure is based on our main analysis result, Theorem 9.1. This procedure yields neural networks with symmetric interconnecting structure with no other constraints on the structure (such as sparsity, which will be addressed in the next chapter). This procedure constitutes an adaptation of the *eigenstructure method* [59], [87] to system (9.1).

Finally, we demonstrated the applicability of our analysis and synthesis results by a specific example (Section 9.5).

CHAPTER 10

SPARSELY INTERCONNECTED NEURAL NETWORKS FOR ASSOCIATIVE MEMORIES WITH APPLICATIONS TO CELLULAR NEURAL NETWORKS

10.1 Introduction

Hardware implementation is one of the final goals in the development of artificial neural networks. Implementations of multilayered feedforward neural networks have a long history of success [119], while implementations of single-layered feedback neural networks seem less successful. Among the various difficulties encountered in VLSI implementations of feedback neural networks, the realization of large numbers of interconnections in networks poses extreme difficulties. In many cases, the large number of interconnections makes the implementations of fully interconnected (feedback) neural networks nearly impossible. Therefore, the reduction in the number of connections is of great practical interest.

In this chapter, we consider implementations of associative memories via artificial neural networks with sparse interconnecting structures. Most of the existing synthesis procedures for associative memories [30], [34], [44], [57], [58], [59], [83], [84], [87], [95], [96], [97], [121], [122] are developed for fully interconnected neural networks and none of them result in neural networks with *prespecified* partial or sparse interconnection structure. Synthesis procedures for neural networks with arbitrarily (prespecified) sparse interconnection structure, or equivalently, with sparse coefficient matrices (having many prespecified zero elements at given locations) constitute a major addition to the development of neural network theory, and such procedures will have potentially many practical applications, especially in the areas of associative memories and pattern recognition (we will define the exact meaning of sparse coefficient matrix later).

We note that if in neural network (9.1), the n neurons are arranged in an array of M rows and N columns, where $n = M \times N$, and if we consider only local interconnections, then (9.1) reduces to a two-dimensional *cellular neural network* (cf. Sections 8.4 and 10.3). In this chapter, we consider a more general case. Specifically, we consider system (9.1) with *arbitrary* interconnections which includes *fully interconnected* nets and cellular neural networks as special cases.

The synthesis techniques developed in Chapter 9 will result in neural networks (9.1) with *symmetric* and *non-sparse* coefficient matrix T. It has been argued by some workers [5], [14], [18], [34], [42], [80], [82], [83], [101], [121] that symmetric interconnections in artificial neural networks are not necessarily always desirable. Moreover, fully interconnected artificial neural networks with even a moderate number of neurons will give rise to large numbers of *line-crossings* resulting from the network interconnections, and thus pose formidable obstacles in (analog) VLSI implementations. For these reasons, it is desirable to develop synthesis procedures which will result in neural networks with an interconnecting structure which does not require symmetry and which does not demand large numbers of interconnections.

Existing work dealing with sparsely interconnected neural networks has been reported in [56]. In this work, a possible solution of transforming a given neural network into a partially connected or cellular network is presented for the discrete-time Hopfield model. However, as pointed out by the author of [56], "the application of the suggested transformation algorithm is severely limited by its quickly growing complexity."

Using the results of Chapter 9, we will develop in Section 10.2 synthesis procedures for neural networks with sparse coefficient matrices in which the interconnection structure is predetermined. We will develop this synthesis procedure for neural network (9.1); however, our method is also applicable to other types of neural network models, such as neural networks defined on hypercubes [59], [87] (cf. Equations (8.4) and (8.5)) and the Hopfield model [44], [45] (cf. Equations (8.1) and (8.3)). In Section 10.3, we apply the sparse synthesis technique developed in Section 10.2 to the design of a class of (nonsymmetric) cellular neural networks. We will show that under certain restrictions on the interconnection structure, system (9.1) is equivalent to a class of zero-input, nonsymmetric cellular neural networks. In Sections 10.4 and 10.5, we consider several specific examples to demonstrate the applicability of our synthesis procedures. Special emphasis is placed on cellular neural networks and networks with different sparse interconnection structures. We conclude with several pertinent remarks in Section 10.6.

10.2 Sparse Design Procedures

Using the results of Chapter 9, we develop in the following design procedures for artificial neural networks which will result in less interconnections and few line-crossings or no line-crossings at all in the interconnections, and which do not require that the interconnection matrix be symmetric. Cellular neural networks, which we address in the next section (with applications to associative memories), are special cases of such *sparsely interconnected artificial neural networks*.

We begin by introducing some necessary terminology.

A matrix $S = [S_{ij}] \in R^{n \times n}$ is said to be an *index matrix,* if it satisfies $S_{ij} = 1$ or 0. The restriction of a matrix $W = [W_{ij}] \in R^{n \times n}$ to an index matrix S, denoted by $W|S$, is defined by $W|S = [h_{ij}]$, where

$$h_{ij} = \begin{cases} W_{ij}, & \text{if } S_{ij} = 1 \\ 0, & \text{otherwise} \end{cases}.$$

We will say that (9.1) is a *neural network with a sparse coefficient matrix* if $T = T|S$ for some given index matrix S.

Sparse Design Problem: Given an $n \times n$ index matrix $S = [S_{ij}]$ with $S_{ii} \neq 0$ for $i = 1, \cdots, n$, and m vectors $\alpha^1, \cdots, \alpha^m$ in B^n, choose $\{A, T, I\}$ with $T = T|S$ in such a manner that $\alpha^1, \cdots, \alpha^m$ are memory vectors of system (9.1).   ∎

Remark 10.1 In the literature, an $n \times n$ matrix is said to be *sparse,* if its number of nonzero elements $<< n^2$. The definition of *sparse coefficient matrix* given in the present chapter is more general and includes the usual definition of sparse matrix as a special case. The sparse design problem considered herein, is in fact, more appropriately called an *indexed design problem.* We will use the term *sparse design* in the present chapter since we wish to be able to make comparisons with the fully connected case.   ∎

A solution for the above sparse design problem is as follows.

Sparse Design Procedure 10.1 Suppose we are given an $n \times n$ index matrix $S =$

$[S_{ij}]$ with $S_{ii} \neq 0$ for $i = 1, \cdots, n$, and m vectors $\alpha^1, \cdots, \alpha^m$ in B^n which are to be stored as memory vectors for (9.1). We proceed as follows:

1) Choose matrix A as the identity matrix.

2) Choose a real number $\mu > 1$ and m vectors $\beta^1, \cdots, \beta^m$, such that $\beta^i = \mu\alpha^i$.

3) Compute the $n \times (m-1)$ matrices

$$Y = [y^1, \cdots, y^{m-1}] = [\alpha^1 - \alpha^m, \cdots, \alpha^{m-1} - \alpha^m],$$

and

$$Z = [z^1, \cdots, z^{m-1}] = [\beta^1 - \beta^m, \cdots, \beta^{m-1} - \beta^m].$$

We let

$$y^i = [y_1^i, \cdots, y_n^i]^T$$

and

$$z^i = [z_1^i, \cdots, z_n^i]^T$$

for $i = 1, \cdots, m-1$.

4) Denote the ith row of the index matrix S by $S_i = [S_{i1}, \cdots, S_{in}]$. For each $i = 1, \cdots, n$, construct two sets M_i and N_i, such that

$$M_i \cup N_i = \{1, \cdots, n\},$$

$$M_i \cap N_i = \phi,$$

and

$$S_{ij} = \begin{cases} 1, & if \ j \in M_i \\ 0, & if \ j \in N_i \end{cases}.$$

Let

$$M_i = \{\sigma_i(1), \cdots, \sigma_i(m_i)\},$$

where $m_i = \sum_{j=1}^n S_{ij}$ and $\sigma_i(k) < \sigma_i(l)$ if $1 \leq k < l \leq m_i$. (Note that m_i is the number of nonzero elements in the ith row of matrix S.)

5) For $i = 1, \cdots, n$, and $l = 1, \cdots, m-1$, let

$$y_{Ii}^l = [y_{\sigma(1)}^l, \cdots, y_{\sigma(m_i)}^l]^T.$$

6) For $i = 1, \cdots, n$, compute the $m_i \times (m-1)$ matrices

$$Y_i = [y_{Ii}^1, \cdots, y_{Ii}^{m-1}],$$

and the $1 \times (m-1)$ vectors

$$Z_i = [z_i^1, \cdots, z_i^{m-1}].$$

7) For $i = 1, \cdots, n$, perform singular value decompositions of Y_i, and obtain

$$Y_i = [U_{i1} \vdots U_{i2}] \begin{bmatrix} D_i & \vdots & 0 \\ \cdots & \vdots & \cdots \\ 0 & \vdots & 0 \end{bmatrix} \begin{bmatrix} V_{i1}^T \\ V_{i2}^T \end{bmatrix},$$

where $D_i \in R^{p_i \times p_i}$ is a diagonal matrix with the nonzero singular values of Y_i on its diagonal and $p_i = \text{rank}(Y_i)$. (Note that $Y_i = U_{i1} D_i V_{i1}^T$.)

8) Compute for $i = 1, \cdots, n$,

$$G_i = [G_{i1}, \cdots, G_{im_i}] = Z_i V_{i1} D_i^{-1} U_{i1}^T + W_i U_{i2}^T,$$

where W_i is an arbitrary $1 \times (m_i - p_i)$ real vector.

9) The matrix $T = [T_{ij}]$ is computed as follows:

$$T_{ij} = \begin{cases} 0, & if \ S_{ij} = 0 \\ G_{ik}, & if \ S_{ij} \neq 0 \ and \ if \ j = \sigma_i(k) \end{cases}. \tag{10.1}$$

10) The bias vector $I = [I_1, \cdots, I_n]^T$ is computed by

$$I_i = \beta_i^m - T_i \alpha^m$$

for $i = 1, \cdots, n$, where T_i is the ith row of T.

Then, $\alpha^1, \cdots, \alpha^m$ will be stored as memory vectors for system (9.1) with A, T, and I determined as above. The states β^i corresponding to α^i, $i = 1, \cdots, m$, will be asymptotically stable equilibrium points of the synthesized system. ∎

Remark 10.2 If in the above synthesis procedure, we choose $A = \text{diag}[a_1, \cdots, a_n]$ with $a_i > 0$, we need to change Z_i in step 6 to $Z_i = [a_i z_i^1, \cdots, a_i z_i^{m-1}]$ and I_i in step 10 to $I_i = a_i \beta_i^m - T_i \alpha^m$. ∎

Remark 10.3 If in the index matrix S there are $q > 1$ identical rows, we can design the corresponding q rows of matrix T simultaneously. For such cases, we need to alter slightly steps 5–9 above. We will demonstrate this idea by means of an example in Section 10.4. We will also demonstrate in Section 10.4 that by special choices of the index matrix S, the Sparse Design Procedure 10.1 can result in a network with few line-crossings, or with no line-crossings at all. ∎

Our next result addresses the existence of a solution for the sparse design problem and the validity of the above design procedure.

Theorem 10.1

1) Solutions for the sparse design problem always exist if $S_{ii} = 1$ for $i = 1, \cdots, n$.

2) The Sparse Design Procedure 10.1 guarantees that $T = T|S$.

3) The Sparse Design Procedure 10.1 guarantees that all vectors in $L_\alpha \cap B^n$, including $\alpha^1, \cdots, \alpha^m$, are stored as memory vectors of system (9.1), where $L_\alpha = \text{Aspan}(\alpha^1, \cdots, \alpha^m)$.

4) The Sparse Design Procedure 10.1 can be applied to any set of desired memory patterns $\alpha^1, \cdots, \alpha^m \in B^n$.

In order for the synthesized system to be a solution of the sparse design problem, we need $G_i Y_i = Z_i$ (which is equivalent to $TY = Z = \mu Y$ and which guarantees T to be a solution for the design problem, cf. Section 9.4) in steps 7 and 8 of the sparse design procedure. Thus, G_i in step 8 is a solution for the sparse design procedure *if and only if*

$$\text{rank}[Y_i] = \text{rank} \begin{bmatrix} Y_i \\ \cdots \\ Z_i \end{bmatrix}.$$

This condition is satisfied if $S_{ii} = 1$, $i = 1, \cdots, n$, since under these conditions, Z_i becomes a row vector which is one of the rows in Y_i multiplied by μ. (This argument is also true if we choose $A = \text{diag}[a_1, \cdots, a_n]$ with $a_i > 0$.) This proves part 1 of the theorem.

Part 2 is clear from (10.1).

To prove part 3, we first check the equilibrium conditions for $\alpha^1, \cdots, \alpha^m$, in which case we require that

$$T\alpha^l + I = A\beta^l = \beta^l, \quad \text{for } l = 1, \cdots, m,$$

where $\beta^l = \mu\alpha^l$ and $\mu > 1$ (therefore, $\beta^l \in (C(\alpha^l))^0$). Using the notation given in the design procedure, we write for $l = 1, \cdots, m - 1$, $y_{Ii}^l = Y_i e_l$, where $e_l \in R^{m-1}$ is a column vector with all elements zero except the lth element which is 1 (cf. step 6 of the sparse design procedure). Also, we have for $i = 1, \cdots, n$,

$$U_{i2}^T y_{Ii}^l = 0 \quad \text{for } l = 1, \cdots, m - 1,$$

and

$$Z_i V_{i1} D_i^{-1} U_{i1}^T Y_i = Z_i.$$

The former is clear from the properties of the singular value decomposition. To see the latter, we recall that $Y_i = U_{i1} D_i V_{i1}^T$ and we assume that Z_i is the jth row of Y_i multiplied by μ (without loss of generality). Then

$$Z_i = \mu \cdot R_j V_{i1}^T$$

where R_j is the jth row of $U_{i1} \times D_i$, and from the properties of the singular value decomposition, we have

$$Z_i V_{i1} D_i^{-1} U_{i1}^T Y_i = \mu \cdot R_j V_{i1}^T \cdot V_{i1} D_i^{-1} U_{i1}^T \cdot U_{i1} D_i V_{i1}^T$$

$$= \mu \cdot R_j V_{i1}^T = Z_i,$$

since $U_{i1}^T U_{i1} = V_{i1}^T V_{i1} = D_i^{-1} D_i = p_i \times p_i$ identity matrix (where p_i is defined in step 7 of the design procedure). According to the design procedure, we compute for $i = 1, \cdots, n$, $T_i y^l = G_i y_{Ii}^l$, and

$$T_i \alpha^l + I_i = T_i y^l + T_i \alpha^m + I_i = G_i y_{Ii}^l + T_i \alpha^m + \beta_i^m - T_i \alpha^m$$

$$= Z_i V_{i1} D_i^{-1} U_{i1}^T Y_i e_l + W_i U_{i2}^T y_{Ii}^l + \beta_i^m = Z_i e_l + \beta_i^m = z_i^l + \beta_i^m$$

$$= \beta_i^l - \beta_i^m + \beta_i^m = \beta_i^l.$$

Hence, $T\alpha^l + I = \beta^l$ for $l = 1, \cdots, m - 1$. For $l = m$, $T\alpha^m + I = \beta^m$ is clear from step 10. Next, we note that the above results imply that

$$T(\alpha^l - \alpha^m) = \beta^l - \beta^m \text{ for } l = 1, \cdots, m - 1,$$

and that for every vector $\alpha \in L_\alpha \cap B^n$, there is a $\lambda = [\lambda_1, \cdots, \lambda_{m-1}] \in R^{m-1}$ such that

$$\alpha = \lambda_1 (\alpha^1 - \alpha^m) + \cdots + \lambda_{m-1} (\alpha^{m-1} - \alpha^m) + \alpha^m.$$

Thus, for every vector $\alpha \in L_\alpha \cap B^n$, we have

$$T\alpha + I = T[\lambda_1 (\alpha^1 - \alpha^m) + \cdots + \lambda_{m-1} (\alpha^{m-1} - \alpha^m)] + T\alpha^m + I$$

$$= \lambda_1 (\beta^1 - \beta^m) + \cdots + \lambda_{m-1} (\beta^{m-1} - \beta^m) + \beta^m$$

$$= \mu[\lambda_1 (\alpha^1 - \alpha^m) + \cdots + \lambda_{m-1} (\alpha^{m-1} - \alpha^m) + \alpha^m]$$

$$= \mu\alpha \overset{\Delta}{=} \beta.$$

Clearly, $\beta \in (C(\alpha))^0$ since $\mu > 1$. By Theorem 9.1 (Corollary 9.4), we see that the states β^i corresponding to α^i, $i = 1, \cdots, m$, as well as the states which correspond to the output vectors in $L_\alpha \cap B^n$ other than α^i, will be asymptotically stable equilibrium points of system (9.1). Therefore, all vectors in $L_\alpha \cap B^n$, including $\alpha^1, \cdots, \alpha^m$, will be stored as memory vectors for system (9.1).

Part 4 follows from similar arguments as in Remark 9.9.   ∎

Remark 10.1 It is emphasized that part 1 of Theorem 10.1 does not imply or require that S_{ij}, $i \neq j$, be zero.   ∎

Remark 10.4 If we wish that the above procedure would result in a system of form (9.1) with $I = 0$, we can modify the Sparse Design Procedure 10.1 as follows:

a) In step 3, let $Y = [\alpha^1, \cdots, \alpha^m]$ and $Z = [\beta^1, \cdots, \beta^m]$.

b) In step 10, let $I = 0$.

Then all conclusions will remain unchanged. In particular, all vectors in

$$\text{Span}(\alpha^1, \cdots, \alpha^m) \cap B^n$$

including $\pm\alpha^1, \cdots, \pm\alpha^m$ are stored as memory vectors of system (9.1).   ∎

10.3 Applications to Cellular Neural Networks

Cellular neural networks, introduced in [22], have found several successful applications in image processing and pattern recognition (see, for example, [23], [76], [77], [78], [79], [124]). On the other hand, applications of (discrete-time) cellular neural networks to associative memories [124], utilizing the outer product method (the Hebbian rule), seem to have been somewhat less successful. As is well known, the outer product method does not guarantee that every desired memory pattern be stored as an equilibrium point (memory point) of the synthesized system when the desired patterns are not mutually orthogonal. Moreover, the storage capacity of networks designed by the outer product method is known to be exceptionally low. Our attempts in solving the design problem of associative memories via cellular neural networks have been successful so far (cf. [63], [70]).

The recent emerging interests in cellular neural networks arise from the fact that the cellular neural networks are among the easiest for VLSI implementations because they use only local interconnections. In the present section, we employ the sparse synthesis techniques developed in the previous section in the design of cellular neural networks with applications to associative memories.

A special class of *two-dimensional cellular neural networks* is described by ordinary differential equations of the form (see [22])

$$\begin{cases} \dot{x}_{ij} = -a_{ij}x_{ij} + \sum_{C(k,l)\in N_r(i,j)} T_{ij,kl}\,\mathrm{sat}(x_{kl}) + I_{ij} \\ y_{ij} = \mathrm{sat}(x_{ij}) \end{cases} \tag{10.2}$$

where $1 \le i \le M$, $1 \le j \le N$, $a_{ij} > 0$, and x_{ij} and y_{ij} are the states and the outputs of the network, respectively.

The basic unit in a cellular neural network is called a *cell*. In (10.2), there are $M \times N$ such cells arranged in an $M \times N$ array. The cell in the ith row and the jth column is denoted by $C(i,j)$, and an *r-neighborhood* $N_r(i,j)$ of the cell $C(i,j)$ for a positive integer r is defined by

$$N_r(i,j) \triangleq \{C(k,l)\colon \max\{|k-i|,|l-j|\} \le r,\ 1 \le k \le M,\ 1 \le l \le N\}.$$

Remark 10.2 System (10.2) characterizes a special class of continuous-time cellular neural networks with square grids, piecewise linear processors, and memoryless interactions (see, e.g., [19]). ∎

Remark 10.5 In the original cellular neural network model introduced by Chua and Yang [22],

$$a_{ij} = \frac{1}{R_x C}$$

(R_x and C are constants), $I_{ij} = I_c$ (I_c is a constant), and $T_{ij,kl} = T_{kl,ij}$. We will not make these assumptions in (10.2). In the applications of cellular neural networks to image processing and pattern recognition, in addition to the bias terms $I_{ij} = I_c$, one frequently requires nonzero input terms

$$\sum_{C(k,l)\in N_r(i,j)} B_{ij,kl}\,u_{kl}$$

in the first equation of (10.2), where u_{ij} represents the inputs of the network. In the present application, we consider zero inputs ($u_{ij} \equiv 0$ for all i and j) and a constant *bias vector* $I = [I_{11}, I_{12}, \cdots, I_{MN}]^T$. This renders (10.2) equivalent to a *nonsymmetric* cellular neural network model by setting the inputs in the original model equal to constants (cf. [18] and [22]). Under these circumstances, we will refer to (10.2) as a (zero-input) *nonsymmetric cellular neural network*. ∎

Using the above nomenclature, we choose a matrix $Q = [Q_{ij,kl}] \in R^{MN \times MN}$ as

$$Q_{ij,kl} = \begin{cases} 1, & \text{if } C(k,l) \in N_r(i,j) \\ 0, & \text{otherwise} \end{cases}. \tag{10.3}$$

With this notation, we see that in order for (9.1) to be equivalent to the nonsymmetric cellular neural network model (10.2), we require in (9.1) $T = T|S$, where $S = Q = [S_{ij}] \in R^{n \times n}$ with $n = M \times N$, and Q is defined in (10.3). Thus, the cellular neural network model (10.2) is a special case of the neural network model (9.1) where the n neurons are arranged in an $M \times N$ array (if $n = M \times N$) and the interconnection structure is confined to local neighborhoods of radius r. Hence, Theorem 9.1 can be applied to analyze the cellular neural network (10.2). Clearly, $S_{ii} = 1$ ($Q_{ij,ij} = 1$), since $C(i,j) \in N_r(i,j)$ for any positive integer r. Thus, the Sparse Design Procedure 10.1 and its modified version (cf. Remark 10.4) can be applied to the design of cellular neural network (10.2) based on the index matrix $S = Q$, where Q is determined in (10.3). By Theorem 10.1, for any given integer $r > 0$ and any set of m vectors $\alpha^1, \cdots, \alpha^m \in B^{MN}$, we can always design a cellular neural network (10.2) so that (10.2) will store $\alpha^1, \cdots, \alpha^m$ as memory vectors.

10.4 Examples

To demonstrate the applicability of the analysis and synthesis procedures presented in the preceding sections, we consider several specific examples.

Example 10.1 We use the Sparse Design Procedure 10.1 to synthesize a cellular neural network (10.2) (or (9.1)) with $n = 12$ ($M = 4$, $N = 3$), and $r = 1$ (neighborhood radius). Given are $m = 4$ vectors specified by

$$\alpha^1 = [1, -1, 1, 1, -1, 1, 1, 1, 1, -1, -1, 1]^T,$$

$$\alpha^2 = [1, 1, 1, 1, -1, -1, 1, -1, -1, 1, 1, 1]^T,$$

$$\alpha^3 = [1, -1, 1, 1, -1, 1, 1, -1, 1, 1, 1, 1]^T,$$

and

$$\alpha^4 = [1, -1, -1, 1, -1, -1, 1, -1, -1, 1, 1, 1]^T.$$

It is desired that these vectors be stored as memory vectors of system (10.2).

We determine the index matrix $S = Q = [S_{ij}] \in R^{12 \times 12}$, where $Q = [Q_{ij,kl}] \in R^{(4 \times 3) \times (4 \times 3)}$ is defined in (10.3). Using the Sparse Design Procedure 10.1, we determine A as the 12×12 identity matrix,

$$
T = \begin{bmatrix}
-1.0000e+01 & 0.0000e+00 & 0 & -1.0000e+01 & -1.0000e+01 & 0 \\
-1.0000e+01 & -2.0000e+00 & 4.0000e+00 & -1.0000e+01 & -1.0000e+01 & -4.0000e+00 \\
0 & -2.6667e+00 & 4.6667e+00 & 0 & -1.0000e+01 & -2.6667e+00 \\
-1.0000e+01 & 0.0000e+00 & 0 & -1.0000e+01 & -1.0000e+01 & 0 \\
-1.0000e+01 & -4.0000e+00 & 4.0000e+00 & -1.0000e+01 & -1.0000e+01 & -2.0000e+00 \\
0 & -4.4000e+00 & 4.4000e+00 & 0 & -1.0000e+01 & -1.2000e+00 \\
0 & 0 & 0 & -1.0000e+01 & -1.0000e+01 & 0 \\
0 & 0 & 0 & -1.0000e+01 & -1.0000e+01 & -4.4409e-16 \\
0 & 0 & 0 & 0 & -1.0000e+01 & 1.0000e+00 \\
0 & 0 & 0 & 0 & 0 & 0 \\
0 & 0 & 0 & 0 & 0 & 0 \\
0 & 0 & 0 & 0 & 0 & 0
\end{bmatrix}
$$

$$
\begin{bmatrix}
0 & 0 & 0 & 0 & 0 & 0 \\
0 & 0 & 0 & 0 & 0 & 0 \\
0 & 0 & 0 & 0 & 0 & 0 \\
-1.0000e+01 & 0.0000e+00 & 0 & 0 & 0 & 0 \\
-1.0000e+01 & -1.0888e-14 & -2.0000e+00 & 0 & 0 & 0 \\
0 & -2.1826e-15 & -1.2000e+00 & 0 & 0 & 0 \\
-1.0000e+01 & -1.3333e+01 & 0 & -6.6667e+00 & -6.6667e+00 & 0 \\
-1.0000e+01 & -1.2667e+01 & -1.3323e-15 & -7.3333e+00 & -7.3333e+00 & -1.0000e+01 \\
0 & -1.000e+01 & 1.000e+00 & 0 & -1.000e+01 & -1.0000e+01 \\
-1.0000e+01 & -1.4000e+01 & 0 & -6.0000e+00 & -6.0000e+00 & 0 \\
-1.0000e+01 & -1.4000e+01 & -2.2538e-15 & -6.0000e+00 & -6.0000e+00 & -1.0000e+01 \\
0 & -1.000e+01 & -7.8505e-16 & 0 & -1.000e+01 & -1.0000e+01
\end{bmatrix},
$$

and

$$
I = [1.2e+01, \ 6.0e+00, \ -1.2667e+01, \ 2.2e+01, \ 1.4e+01, \ -1.44e+01, \ 1.2e+01,
$$

$$
2.0e+01, \ -8.8818e-15, \ 1.0e+01, \ 2.0e+01, \ 1.2e+01]^T.
$$

(In the above computations, we chose $\mu = 2$ in step 2 and chose $W_i = -10 \times O_{m_i} \times U_{i2}$ in step 8 of the Sparse Design Procedure 10.1, where $O_{m_i} = [1, \cdots, 1] \in R^{1 \times m_i}$ and $m_i = \sum_{j=1}^{n} S_{ij}$.)

Using Theorem 9.1, we verified that $\alpha^1, \cdots, \alpha^4$ are memory vectors of the synthesized system (10.2) with $\{A, T, I\}$ given above.

By Theorem 9.1, we determine that system (10.2) has 2 additional memory vectors in B^n given by

$$\alpha^5 = [1, -1, -1, 1, -1, -1, 1, 1, -1, -1, -1, 1]^T,$$

and

$$\alpha^6 = [1, 1, 1, 1, -1, -1, 1, 1, -1, -1, -1, 1]^T.$$

Their corresponding asymptotically stable states are given by $\beta^5 = 2\alpha^5$ and $\beta^6 = 2\alpha^6$, respectively. System (10.2) has 1 additional memory vector in $D^n - B^n$, given by

$$\alpha^7 = [-0.7273, -1, -1, 1, 1, -1, -0.7273, -1, -1, 1, 1, 1]^T.$$

Its corresponding asymptotically stable state is given by

$$\beta^7 = [-0.7273, -4.7273, -22, 16.5455, 12.5455, -22, -0.7273, -4.7273,$$

$$-22, 19.2727, 19.2727, 2]^T.$$

System (10.2) has also 5 unstable equilibrium points. ∎

Example 10.2 In order to ascertain how typical the results of Example 10.1 are, we repeat the example twenty times using different sets of desired vectors to be stored as memory vectors. Each set contained $m = 4$ vectors in B^n which are generated randomly. For each given set of vectors, we synthesized system (10.2) using the Sparse Design Procedure 10.1. Table 10.1 summarizes our findings. Also shown in Table 10.1 are the results for system (9.1) synthesized by Synthesis Procedure 9.1 (which usually results in fully connected neutal networks) *for the same sets of desired vectors* to be stored as memories.

From Table 10.1, we see that in the present example, the cellular neural network implementations for associative memories have more spurious states than the fully connected networks. Also, there are more unstable equilibrium points in the synthesized cellular neural networks than in the fully connected networks in the present example. ∎

Example 10.3 We now present several problems which to the best of our knowledge cannot be addressed by other synthesis procedures for associative memories. In all cases, we consider a neural network with 16 neurons ($n = 16$) and in all cases our objective is to store

	The cellular neural network with $r = 1$	The fully connected neural network
average of total number of memory vectors in B^n	12.2	4.5
average of total number of undesired memory vectors in B^n	8.2	0.5
average of total number of memory vectors in $D^n - B^n$	3.1	0.4
average of total number of unstable equilibrium points in R^n	64.65	14.25
total number of desired patterns which were not stored as memory vectors	0	0

Table 10.1: Experimental results for Example 10.2

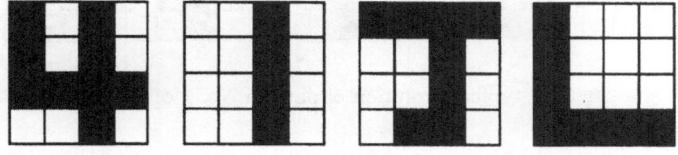

Figure 10.1: The four desired memory patterns used in Example 10.3

the four patterns shown in Figure 10.1 as memories. As indicated in this figure, sixteen boxes are used to represent each pattern (in R^{16}), with each box corresponding to a vector component which is allowed to assume values from -1 to 1. For purpose of visualization, -1 will represent white, 1 will represent black, and the intermediate values will correspond to appropriate grey levels, as shown in Figure 10.2.

The four cases which we consider below, are synthesized by the Sparse Design Procedure 10.1. These cases involve different prespecified constraints on the interconnecting structure of each network.

Case I: Cellular neural network. We design a cellular neural network (10.2) with $r = 1$, $M = 4$, and $N = 4$. (Due to space limitations, we will not display the interconnecting matrix T for the present case, as well as for the three subsequent cases.) The performance

144

-1   -0.5   0   0.5   1

Figure 10.2: Grey levels

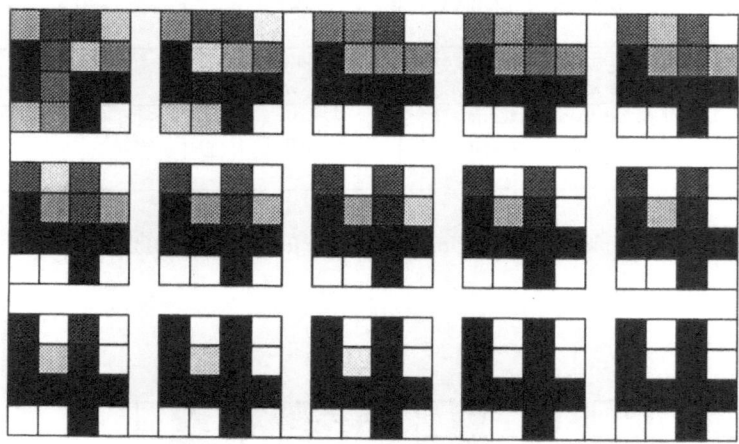

Figure 10.3: A typical evolution of pattern No. 1 of Figure 10.1

of this network is illustrated by means of a typical simulation run of equation (10.2) (or equivalently, equation (9.1)) using equation(9.2), shown in Figure 10.3. In this figure, the desired memory pattern is depicted in the lower right corner. The initial state, shown in the upper left corner, is generated by adding to the desired pattern zero-mean Gaussian noise with a standard deviation SD=1. The iteration of the simulation evolves from left to right in each row and from the top row to the bottom row. The desired pattern is recovered in 14 steps with a step size $h = 0.2$ in the digital simulation of equation (9.1). We do not identify a unit for the step size h. In view of Remark 9.6, the unit could be seconds, milliseconds, or any other small time intervals. All simulations for the present chapter are performed on a Sun SPARC Station using MATLAB.

Case II: Reduction of line-crossings. We arrange the 16 neurons in a 4×4 array and we consider only horizontal and vertical interconnections. For this case, the index matrix

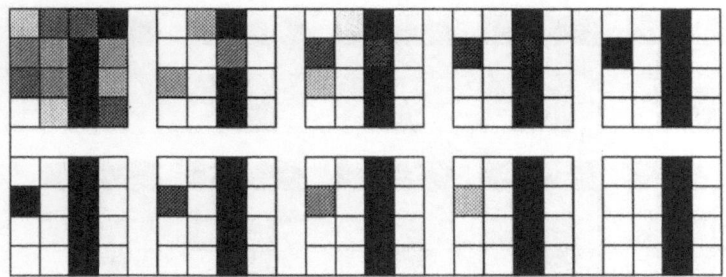

Figure 10.4: A typical evolution of pattern No. 2 of Figure 10.1

$S = Q = [S_{ij}] \in R^{16 \times 16}$, and $Q = [Q_{ij,kl}] \in R^{(4 \times 4) \times (4 \times 4)}$ assumes the form

$$Q_{ij,kl} = \begin{cases} 1, & \text{if } i = k \text{ or } j = l \\ 0, & \text{otherwise} \end{cases} \tag{10.4}$$

A typical simulation run for the present case is depicted in Figure 10.4. In this figure, the noisy pattern is generated by adding to the desired pattern uniformly distributed noise defined on $[-0.7, 0.7]$. Convergence occurs in 9 steps with $h = 0.2$.

We emphasize that by choosing the index matrix as in (10.4), we are able to reduce significantly the number of line-crossings, which is of great concern in VLSI implementations of artificial neural networks.

Case III: Two rows of S identical (see Remark 10.3). In this case, we choose an index matrix $S = [S_{ij}] \in R^{16 \times 16}$ of the form

$$S_{ij} = \begin{cases} 1, & \text{if } i = 1 \text{ or } i = 16 \text{ or } j = 1 \text{ or } j = 16 \text{ or } |i - j| \le 1 \\ 0, & \text{otherwise} \end{cases}.$$

This requires that the T matrix has zero elements everywhere except in its first and last rows, its first and last columns, and in its tridiagonal elements.

Rows 2 to 15 are designed by using the Sparse Design Procedure 10.1, step by step. Since the first row S_1 and the last row S_{16} of S are identical, we can design the rows T_1 and T_{16} of T simultaneously. To see this, we take in step 5 of the design procedure $y_{I1}^l = [y_{\sigma(1)}^l, \cdots, y_{\sigma(m_1)}^l]^T$ and $y_{I16}^l = [y_{\sigma(1)}^l, \cdots, y_{\sigma(m_{16})}^l]^T$ for $l = 1, \cdots, m - 1$. Clearly, $m_1 = m_{16} = n = 16$ and $y_{I1}^l = y_{I16}^l$ for $l = 1, \cdots, m - 1$, since $S_1 = S_{16} = [1, \cdots, 1] \in R^{1 \times 16}$. In step

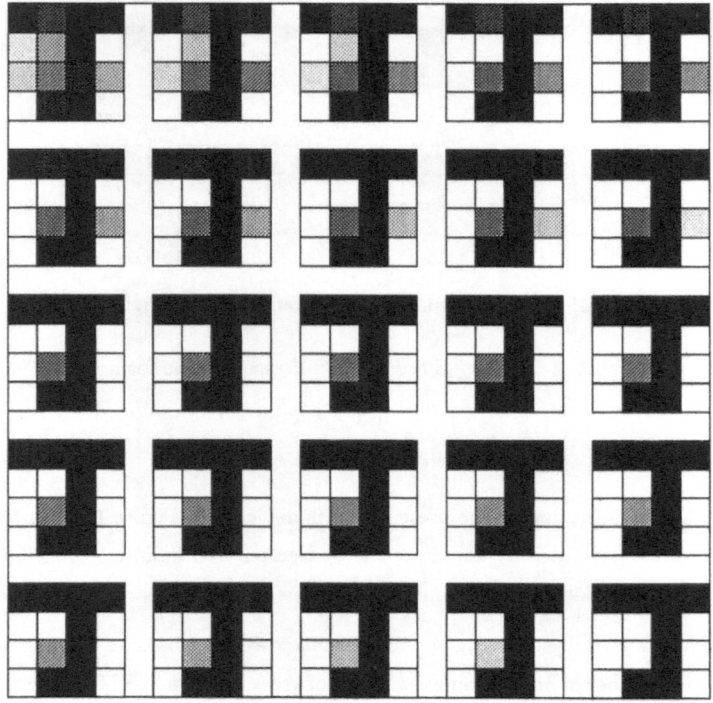

Figure 10.5: A typical evolution of pattern No. 3 of Figure 10.1

6, we take $Y_1 = [y_{I1}^1, \cdots, y_{I1}^{m-1}]$ and the $2 \times (m-1)$ vector

$$Z_1 = \begin{bmatrix} z_1^1 & & z_1^{m-1} \\ & \cdots & \\ z_{16}^1 & & z_{16}^{m-1} \end{bmatrix}.$$

In step 7, we perform a singular value decomposition of Y_1 and obtain U_{11}, U_{12}, D_1, and V_{11}. In step 8, we compute $G_1 = Z_1 V_{11} D_1^{-1} U_{11}^T + W_1 U_{12}^T$, where W_1 is an arbitrary $2 \times (m_1 - p_1)$ real matrix and $p_1 = \text{rank}(Y_1)$. In step 9, we determine T_1 from the first row of G_1 and T_{16} from the second row of G_1, using (10.1).

A typical simulation run for this network is shown in Figure 10.5. In this case, the noisy pattern is generated by adding Gaussian noise $N(0, 0.5)$ to the desired pattern. Convergence occurs in 24 steps with $h = 0.2$.

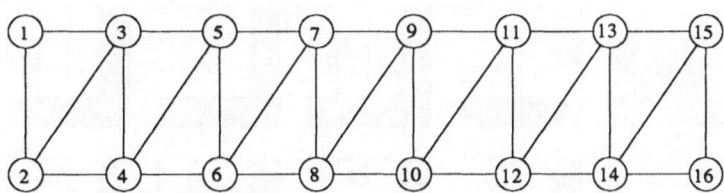

Figure 10.6: A possible structure for a neural network without line-crossings in the interconnecting structure.

We can generalize the above case to design problems for which the index matrix S has several identical rows. In particular, if $S = [S_{ij}]$, $S_{ij} = 1$ for all i and j, the Sparse Design Procedure 10.1 reduces to a procedure for a *fully connected* neural network (9.1), and the reduced design procedure (where all rows of T are determined simultaneously) will be more general than Synthesis Procedure 9.1. To see this, note that the reduced design procedure will generally result in a nonsymmetric T and by special choice of matrix W in step 8, the reduced design procedure will become Synthesis Procedure 9.1.

Case IV: Quinquediagonal matrix S resulting in an interconnecting structure without line-crossings. We choose $S = [S_{ij}] \in R^{16 \times 16}$ as

$$S_{ij} = \begin{cases} 1, & \text{if } |i - j| \leq 2 \\ 0, & \text{otherwise} \end{cases}. \tag{10.5}$$

This will result in a quinquediagonal matrix S, enabling us to arrange the $n = 16$ neurons in the configuration shown in Figure 10.6. Note that in this figure there are no line-crossings. Furthermore, note that this configuration can be generalized to arbitrary n.

A typical simulation run for the present case is depicted in Figure 10.7. In this figure, the noisy pattern is generated by adding Gaussian noise $N(0.1, 0.7)$ to the desired pattern. Convergence occurs in 13 steps with $h = 0.2$. ∎

10.5   Chinese Character Recognition

In [23], cellular neural networks have been proposed for a possible application to the Chinese character recognition. As pointed out in [23], a major drawback of the existing methods for Chinese character recognition using digital computers is the slow recognition

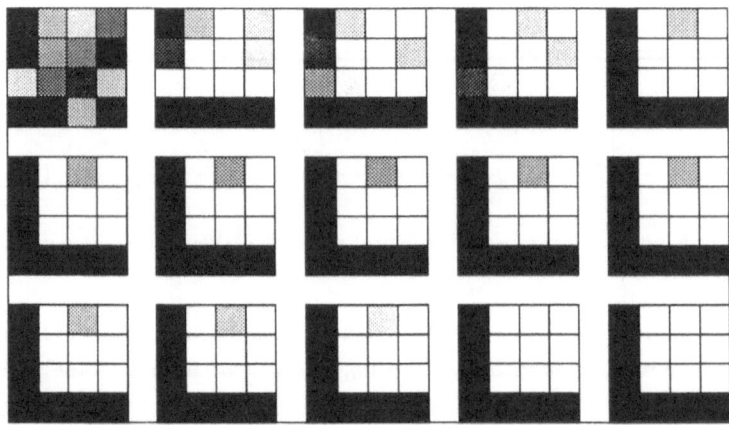

Figure 10.7: A typical evolution of pattern No. 4 of Figure 10.1

speed. Using cellular neural networks as a tool for Chinese character recognition, the recognition speed can be increased significantly because cellular neural networks are usually implemented by the analogue VLSI technology.

In this section, we propose to solve the Chinese character recognition problem using cellular neural networks. Our approach of using cellular neural networks for Chinese character recognition differs significantly from the one proposed in [23]. It is suggested in [23] that cellular neural networks might have the capability of extracting certain features of images (also of Chinese characters) using appropriate dynamic rules (connection matrices). In our method to be presented in the following, we consider the cellular neural network as a model for associative memories, and design a cellular neural network which can remember certain Chinese characters.

To demonstrate our method, we consider a small set of basic Chinese characters as our desired memory patterns which are represented by bipolar vectors. We consider the 25 desired memory patterns $\alpha^1, \cdots, \alpha^{25}$ shown in Figure 10.8. (The upper left pattern is denoted by α^1 and the lower right pattern is denoted by α^{25}.) These patterns constitute modules which represent individually or in combination, Chinese characters. They are coded in R^{81} in a similar manner as was done in Example 10.3 (including usage of grey levels, as described in Example 10.3).

We wish to synthesize a cellular neural network (10.2) with $n = 81$ ($M = N = 9$) and $r = 3$, which will "remember" these modules. As mentioned above, some of the Chinese characters can be represented by two modules. In particular, the patterns given in Figure 10.8 can be used to generate at least 50 commonly used Chinese characters. To demonstrate this, we add one more vector, α^{26}, with every entry equal to 1 (black), to the set of desired memory patterns. In doing so, we can generate desired combinations for Chinese characters which are made up of some of the basic modules given in Figure 10.8. For instance, the character corresponding to α^6 means "sun" and the character corresponding to α^{14} means "moon". A new Chinese character can be generated as $\alpha^{27} = \alpha^6 + \alpha^{14} + \alpha^{26} \in \mathrm{Span}(\alpha^1, \cdots, \alpha^{26}) \cap B^{81}$, which means "bright" (see Figure 10.9). Using the modified Sparse Design Procedure 10.1 as discussed in Remark 10.4, we only need to synthesize a system (9.1), in which the 81 neurons are arranged in a 9×9 array and the interconnections are restricted to local neighborhoods of radius $r = 3$, by employing these basic patterns. The resulting system will automatically "remember" all possible combinations of these basic components, which include the 50 commonly used Chinese characters mentioned above (by Theorem 10.1 and Remark 10.4).

In the modified Sparse Design Procedure 10.1, by taking $\mu = 2$ in step 2 and $W_i = -10 \times O_{m_i} \times U_{i2}$ in step 8, where $O_{m_i} = [1, \cdots, 1] \in R^{1 \times m_i}$ and $m_i = \sum_{j=1}^n S_{ij}$, we design a neural network of the form (9.1) with the above specifications (*i.e.*, a cellular neural network of the form (10.2) with $M = N = 9$ and $r = 3$) which stores $\alpha^1, \cdots, \alpha^{26}$ as memory vectors. This system has 2601 total interconnections, while a fully connected neural network with $n = 81$ will have a total of 6561 interconnections. By using the cellular neural network of the present example, we are able to reduce the total number of required interconnections to less than 40%.

A typical simulation run, involving the pattern $\alpha^{27} = \alpha^6 + \alpha^{14} + \alpha^{26}$ is depicted in Figure 10.10. The noisy initial pattern in Figure 10.10 (upper left corner) is generated by adding to α^{27} zero-mean Gaussian noise with a standard deviation SD=1. The desired pattern α^{27} is recovered in 24 steps with a step size $h = 0.227$ (lower right corner in Figure 10.10).

Simulation results showed that all the other vectors corresponding to the aforementioned 50 commonly used Chinese characters are (reachable) memory vectors of the synthesized cellular neural network.

For the same initial noisy pattern shown in Figure 10.10, the desired pattern is recovered

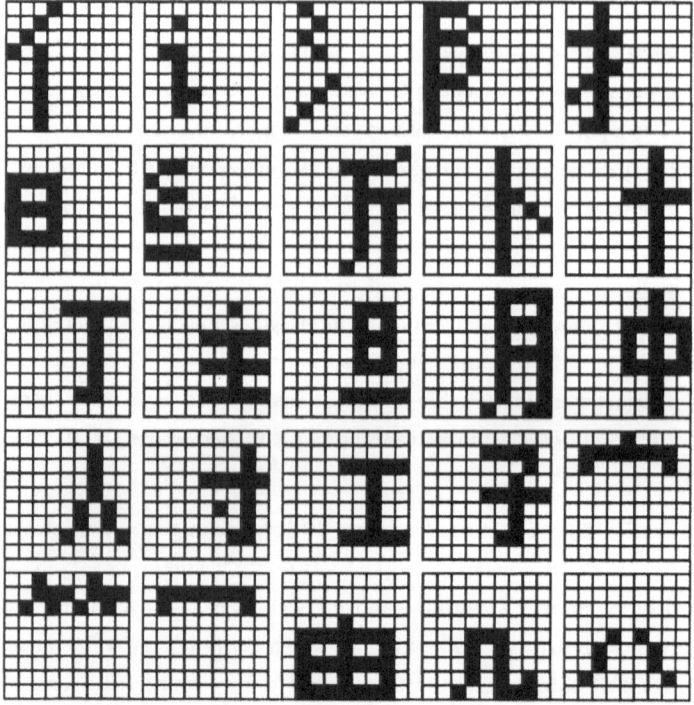

Figure 10.8: The twenty-five desired memory patterns

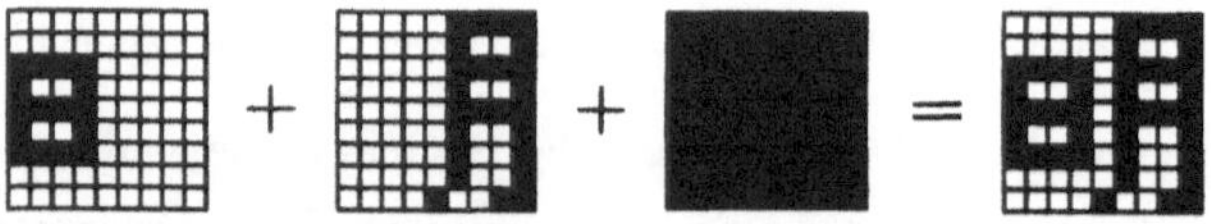

Figure 10.9: The Chinese character composed of patterns No. 6 and No. 14 in Figure 10.8

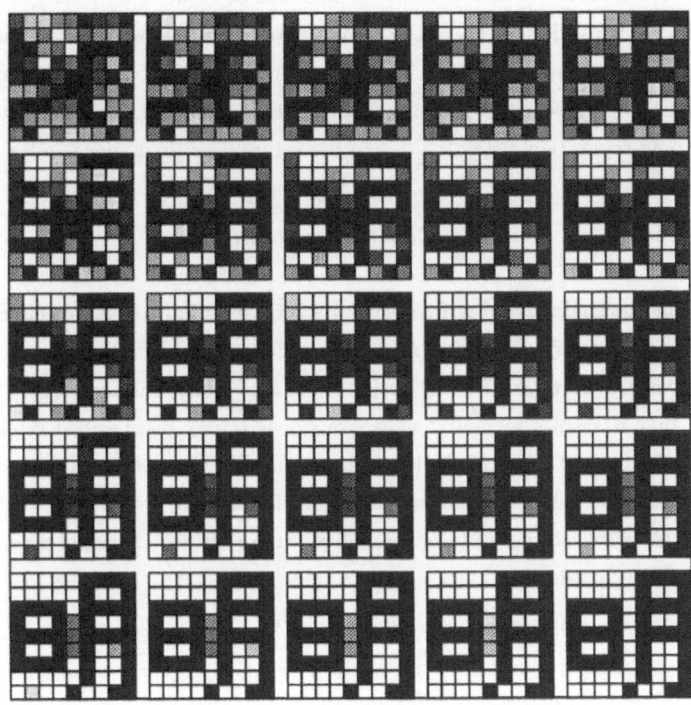

Figure 10.10: A typical evolution of the Chinese character composed of patterns No. 6 and No. 14 in Figure 10.8

in 8 steps, with the same step size, when using a fully connected neural network (9.1) designed by the modified Synthesis Procedure 9.1 (as discussed in Remark 9.8) for the same desired set of memory patterns $\alpha^1, \cdots, \alpha^{26}$. One of the reasons for the lower convergence speed of cellular neural networks is that we only use local interconnections in such systems.

The most commonly used Chinese characters are roughly 6700 in number. More than half of these consist of approximately 500 basic characters (modules) or combinations of these characters, as described above. This example can be expanded by designing a cellular neural network which will store these 500 basic Chinese characters as well as combinations of these characters. In doing so, we will have stored over one half of the 6700 commonly used Chinese characters. The remaining commonly used characters (numbering about 3000), will have to be stored separately, using the design procedure described above.

10.6   Concluding Remarks

By utilizing the analysis and synthesis results of Chapter 9, we developed in Section 10.2 a synthesis procedure for associative memories which are realized by means of sparsely interconnected neural networks (Sparse Design Procedure 10.1). The advantage of our design procedure is that it results in neural networks which satisfy prespecified interconnecting structures. In Section 10.3, we applied this design procedure to the synthesis of a class of cellular neural networks for associative memories. Finally, in Sections 10.4 and 10.5, we demonstrated the applicability and the versatility of the results established herein by means of several specific examples.

The significance of the results presented in this chapter is that we can synthesize by the present methods artificial neural networks which have a *prespecified interconnecting structure* and which *guarantee to store any desired set of bipolar patterns* as memories provided that the interconnecting structure includes self feedback for every neuron. Design procedures which will result in neural networks with prespecified interconnecting structure with no self feedback are left for future investigation.

Our work in the present chapter constitutes the first successful synthesis procedure for associative memories by means of artificial neural networks with *arbitrarily prespecified partial or sparse* interconnecting structure. We believe that the sparse synthesis technique advanced herein will have potentially many applications in associative memories and pattern

153

recognition, since our method can be adapted for many other neural network models (e.g., neural network models described by (8.3)–(8.6)) and since the neural networks synthesized by the present method are among the easiest for analog VLSI implementations.

CHAPTER 11

ROBUSTNESS ANALYSIS OF A CLASS OF SPARSELY INTERCONNECTED NEURAL NETWORKS WITH APPLICATIONS TO THE DESIGN PROBLEM

11.1 Introduction

In the preceding two chapters, we considered neural networks described by equations of the form (cf. (9.1))

$$\begin{cases} \dot{x} = -Ax + T\text{sat}(x) + I \\ y = \text{sat}(x) \end{cases} \tag{11.1}$$

In Chapter 9, we established *analysis* results which enable us to locate all equilibrium points of (11.1) and ascertain their qualitative properties in a systematic manner. Also, in Chapters 9 and 10, we developed *synthesis* procedures for associative memories by means of neural networks (11.1) which guarantee to store any set of *desired* bipolar patterns as memories and which have *predetermined* interconnection structures. The synthesis procedures enable us to synthesize neural networks which are either *fully* interconnected, or *partially* (or *sparsely*) interconnected.

In the present chapter, our focus will be on some of the problems encountered in the hardware implementations of *associative memories* via artificial neural networks modeled by (11.1). We will call a vector α a *memory vector* (or simply, a *memory*) of system (11.1), if $\alpha = \text{sat}(\beta)$ and if $\beta \in R^n$ is an asymptotically stable equilibrium point of system (11.1). In many practical applications, the desired memory patterns are represented by a set of bipolar vectors (or binary vectors). We present in Section 11.2 a robustness analysis of the stability properties of bipolar type memory vectors for neural network (11.1). Specifically, we will assume that $\alpha^1, \cdots, \alpha^m \in B^n \triangleq \{x \in R^n : x_i = 1 \text{ or } -1, i = 1, \cdots, n\}$ are the desired memory vectors of system (11.1) and we will investigate under what conditions $\alpha^1, \cdots, \alpha^m$

are *also* memory vectors of the perturbed system described by

$$
\begin{cases}
\dot{x} = -(A + \Delta A)x + (T + \Delta T)\text{sat}(x) + (I + \Delta I) \\
y = \text{sat}(x)
\end{cases}
\tag{11.2}
$$

where $\Delta A = \text{diag}[\Delta a_1, \cdots, \Delta a_n]$ with $a_i + \Delta a_i > 0$ for $i = 1, \cdots, n$, $\Delta T \in R^{n \times n}$, and $\Delta I \in R^n$. This problem is of great interest from a practical point of view [71], especially in the analog VLSI implementations of system (11.1), since one cannot realize *precisely* the synthesized parameters $\{A, T, I\}$, and one can consider ΔA, ΔT, and ΔI as parameter inaccuracies resulting from the implementation process. We will establish a result which enables us to determine allowable upper bounds for the permissible perturbations ΔA, ΔT, and ΔI in terms of the expression

$$
\|A^{-1}\Delta A\|_\infty + \|A^{-1}\Delta T\|_\infty + \|A^{-1}\Delta I\|_\infty,
$$

where $\|\cdot\|_\infty$ denotes the matrix norm induced by the l_∞ vector norm (see Theorem 11.1).

Utilizing the above results and the results of Chapter 10, we will solve in Section 11.3 the following problem: Given $\alpha^1, \cdots, \alpha^m \in B^n$ as desired patterns and given a prespecified interconnecting structure (in terms of sparsity), find A, T, and I such that $\alpha^1, \cdots, \alpha^m$ become memory vectors of system (11.1), such that the system satisfies the prespecified interconnecting structure, and such that the connection matrix is *symmetric*.

In Section 11.4, we consider several examples to demonstrate the applicability and versatility of the present results, and in Section 11.5, we conclude this chapter with several pertinent remarks.

11.2 Robustness Analysis

In the sequel, we will make use of the notation

$$
\delta(x) = \min_{1 \le i \le n} \{|x_i|\} \quad \text{for} \ x \in R^n
$$

and

$$
E(\alpha) = \{x \in R^n : \ x_i \alpha_i > 1, \ i = 1, \cdots, n\}
$$

for $\alpha \in B^n$. Recall that for $x \in R^n$, the l_∞ vector norm is defined by

$$
\|x\|_\infty = \max_{1 \le i \le n} \{|x_i|\}.
$$

Also recall that the matrix norm induced by the l_∞ vector norm for a matrix $F = [f_{ij}] \in R^{m \times n}$ is defined by

$$\|F\|_\infty = \max_{1 \le i \le m} \Big\{ \sum_{j=1}^n |f_{ij}| \Big\}.$$

We will make use of the following result which has been proven in Chapter 9. (See Corollary 9.4 and notice that $E(\alpha) = (C(\alpha))^0$.)

Lemma 11.1 Let $\alpha \in B^n$. If

$$\beta = A^{-1}(T\alpha + I) \in E(\alpha)$$

then (α, β) is a pair of memory vector and asymptotically stable equilibrium point of system (11.1). ∎

We are now in a position to prove the next result.

Theorem 11.1 Suppose that $\alpha^1, \cdots, \alpha^m \in B^n$ are desired memory vectors of system (11.1), and suppose that $\beta^1, \cdots, \beta^m$ are asymptotically stable equilibrium points of system (11.1) corresponding to $\alpha^1, \cdots, \alpha^m$, respectively. Let

$$\nu = \min_{1 \le l \le m} \{\delta(\beta^l)\}. \tag{11.3}$$

Then $\alpha^1, \cdots, \alpha^m$ are also memory vectors of system (11.2) provided that

$$\|A^{-1}\Delta A\|_\infty + \|A^{-1}\Delta T\|_\infty + \|A^{-1}\Delta I\|_\infty < \nu - 1. \tag{11.4}$$

From Lemma 11.1, we see that, for $l = 1, \cdots, m$,

$$\beta^l = A^{-1}(T\alpha^l + I) \in E(\alpha^l),$$

or equivalently,

$$a_i^{-1}(T_i\alpha^l + I_i) = \beta_i^l \text{ with } |\beta_i^l| > 1 \text{ for } i = 1, \cdots, n, \tag{11.5}$$

where a_i is the ith diagonal element of matrix A, T_i represents the ith row of matrix T, and I_i and β_i^l are the ith element of I and β^l, respectively. In the rest of the proof, we assume that

$$\|A^{-1}\Delta T\|_\infty + \|A^{-1}\Delta I\|_\infty \le \eta \tag{11.6}$$

and

$$\|A^{-1}\Delta A\|_\infty < \nu - \eta - 1 \tag{11.7}$$

i.e., (11.4) is satisfied. We will show that $\alpha^1, \cdots \alpha^m$ are also memory vectors of the neural network described by (11.2).

For $l = 1, \cdots, m$, compute $A^{-1}(\Delta T \alpha^l + \Delta I)$ and apply (11.6) to obtain

$$|a_i^{-1}\Delta T_i \alpha^l + a_i^{-1}\Delta I_i| \leq \sum_{j=1}^{n} |a_i^{-1}\Delta T_{ij}| + \max_{1 \leq i \leq n} |a_i^{-1}\Delta I_i|$$

$$\leq \|A^{-1}\Delta T\|_\infty + \|A^{-1}\Delta I\|_\infty \leq \eta \tag{11.8}$$

where $\Delta T_i = [\Delta T_{i1}, \cdots, \Delta T_{in}]$ represents the ith row of ΔT and ΔI_i is the ith component of ΔI.

We now compute, using (11.5),

$$\bar{\beta}_i^l \triangleq (a_i + \Delta a_i)^{-1}[(T_i + \Delta T_i)\alpha^l + I_i + \Delta I_i]$$

$$= \frac{a_i}{a_i + \Delta a_i} [a_i^{-1}(T_i\alpha^l + I_i) + a_i^{-1}\Delta T_i\alpha^l + a_i^{-1}\Delta I_i]$$

$$= \frac{a_i}{a_i + \Delta a_i} (\beta_i^l + a_i^{-1}\Delta T_i\alpha^l + a_i^{-1}\Delta I_i). \tag{11.9}$$

From (11.3) and (11.8), when $\beta_i^l > 1$ $(\alpha_i^l = 1)$, we have

$$\bar{\beta}_i^l \geq \frac{a_i}{a_i + \Delta a_i} (\beta_i^l - |a_i^{-1}\Delta T_i\alpha^l + a_i^{-1}\Delta I_i|)$$

$$\geq \frac{a_i}{a_i + \Delta a_i}(\nu - \eta) > 1. \tag{11.10}$$

Also, when $\beta_i^l < -1$ $(\alpha_i^l = -1)$, we have

$$\bar{\beta}_i^l \leq \frac{a_i}{a_i + \Delta a_i} (\beta_i^l + |a_i^{-1}\Delta T_i\alpha^l + a_i^{-1}\Delta I_i|)$$

$$\leq \frac{a_i}{a_i + \Delta a_i}(-\nu + \eta) < -1. \tag{11.11}$$

Relations (11.10) and (11.11) are true since (11.7) implies that

$$1 + \frac{\Delta a_i}{a_i} \leq 1 + \left|\frac{\Delta a_i}{a_i}\right| < \nu - \eta$$

which is equivalent to

$$\frac{a_i}{a_i + \Delta a_i} (\nu - \eta) > 1,$$

by noticing that $a_i > 0$ and $a_i + \Delta a_i > 0$. (11.9), (11.10) and (11.11) in turn imply that

$$\bar{\beta} = (A + \Delta A)^{-1}[(T + \Delta T)\alpha^l + (I + \Delta I)] \in E(\alpha^l)$$

for $l = 1, \cdots, m$. From Lemma 11.1, we now see that $\alpha^1, \cdots, \alpha^m$ are also memory vectors for system (11.2). ∎

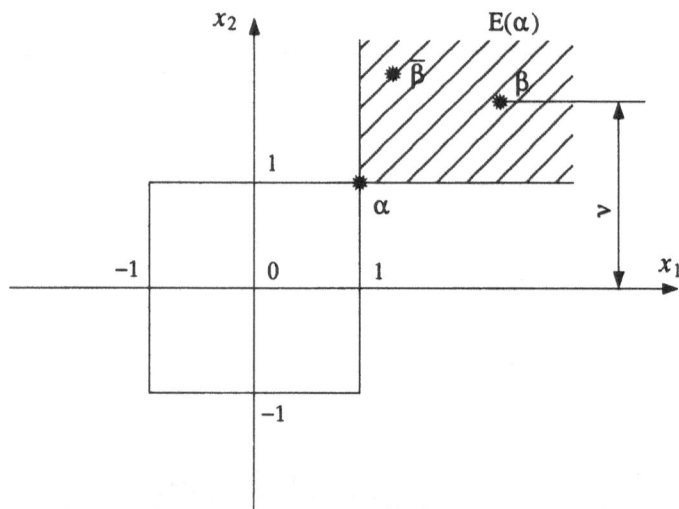

Figure 11.1: A geometric interpretation of Theorem 11.1

In the following, we will give a geometric interpretation of Theorem 11.1 in R^2. Suppose that $\alpha \in R^2$ is a (desired) memory vector of neural network (11.1) and its corresponding asymptotically stable equilibrium point is β. Then, $\beta = A^{-1}(T\alpha + I)$ *must* be in the region given by $E(\alpha)$ (cf. the crosshatched region in Figure 11.1), since we have $\nu > 1$ in Theorem 11.1.

When we have perturbations ΔA, ΔT, and ΔI as in system (11.2), the vector β will be displaced from its original location to, say, $\bar{\beta}$. In order for α to remain as a memory vector for system (11.1) after perturbation (*i.e.*, for α to be a memory vector for system (11.2)), we require that $\bar{\beta}$ also be in $E(\alpha)$. It is clear from Lemma 11.1 that as long as $\bar{\beta}$ is in $E(\alpha)$, α will be a memory vector of the perturbed system (11.2). Theorem 11.1 gives one of the *possible* upper bounds for the perturbations, specified by

$$\|A^{-1}\Delta A\|_\infty + \|A^{-1}\Delta T\|_\infty + \|A^{-1}\Delta I\|_\infty,$$

which will ensure that the perturbed vector $\bar{\beta}$ and the original vector β are within the same region given by $E(\alpha)$. This upper bound is given by $\nu - 1$ (if ν satisfies condition (11.3)).

Remark 11.1 In system (11.2), we have to require that $a_i + \Delta a_i > 0$ for each $i = 1, \cdots, n$. From Lemma 11.1, we see that only perturbations ΔA with $\Delta a_i < 0$ for $i = 1, \cdots, n$ (*i.e.*, $\Delta T = 0$ and $\Delta I = 0$) will not change the desired memory vectors $\alpha^1, \cdots \alpha^m \in B^n$ of system (11.1). ∎

Remark 11.2 When considering perturbations due to an implementation process, the focus is usually on the interconnection matrix T and not on the parameters A and I. Assuming $\Delta A = 0$ and $\Delta I = 0$, system (11.2) takes the form

$$\begin{cases} \dot{x} = -Ax + (T + \Delta T)\text{sat}(x) + I \\ y = \text{sat}(x) \end{cases} \tag{11.12}$$

and condition (11.4) assumes the form

$$\|A^{-1}\Delta T\|_\infty \le \nu - 1. \tag{11.13}$$

∎

11.3   Applications to the Design Problem

In this section, we first present the rationale and a summary for the synthesis procedures developed in Chapters 9 and 10. We then present a synthesis procedure for neural network (11.1) with sparsity and symmetry constraints.

A.   Rationale of the Synthesis Procedure with Sparsity Constraints (non-symmetric interconnection matrix)

Suppose we are given a set of desired patterns $\alpha^1, \cdots, \alpha^m \in B^n$. We wish to design a system of form (11.1) which stores $\alpha^1, \cdots, \alpha^m$ as memories. Without loss of generality, we choose A as the $n \times n$ identity matrix. We denote

$$Y = [\alpha^1 - \alpha^m, \cdots, \alpha^{m-1} - \alpha^m]$$

and choose $\beta^l = \mu \alpha^l$ for $l = 1, \cdots, m$, with $\mu > 1$ (hence, $\beta^l \in E(\alpha^l)$). In view of Lemma 11.1, it can be verified that in order for system (11.1) to store the desired patterns $\alpha^1, \cdots, \alpha^m$ as memories and to store $\beta^1, \cdots, \beta^m$ as corresponding asymptotically stable equilibrium points, matrix T must be a solution of the matrix equation,

$$TY = \mu Y. \tag{11.14}$$

Solutions of (11.14) (for T) always exist since

$$\text{rank}[Y] = \text{rank} \begin{bmatrix} Y \\ \cdots \\ \mu Y \end{bmatrix}.$$

Indeed, a trivial solution for T is the $n \times n$ identity matrix multiplied by μ. We desire to find non-trivial solutions for matrix T. This can be accomplished by many methods, and in our research, we utilize the *singular value decomposition method*. Performing a singular value decomposition of Y, we obtain

$$Y = [U_1 \vdots U_2] \begin{bmatrix} D & \vdots & 0 \\ \cdots & \vdots & \cdots \\ 0 & \vdots & 0 \end{bmatrix} \begin{bmatrix} V_1^T \\ \cdots \\ V_2^T \end{bmatrix}, \tag{11.15}$$

where $D \in R^{p \times p}$ is a diagonal matrix with the nonzero singular values of matrix Y on its diagonal and

$$p = \text{rank}[Y]. \tag{11.16}$$

Then, (11.14) yields

$$TU_1 = \mu U_1. \tag{11.17}$$

Solutions of (11.17) for T can be expressed as

$$T = \mu U_1 U_1^T + W U_2^T \tag{11.18}$$

where W is an arbitrary $n \times (n - p)$ real matrix. It can easily be verified that T given in (11.18) is also a solution of (11.14) and it is non-trivial when $p < n$. The above observations constitute a mathematical basis for the design procedure of neural networks with *no constraints* on the interconnecting structure. This procedure usually results in a *fully interconnected* neural network (for a special treatment, refer to Chapter 9).

We next consider the case in which we have *constraints on the interconnecting structure*. Specifically, we will consider constraints which require that predetermined elements of matrix T be zero. To simplify the subsequent discussion, we consider the specific case when $n = 4$ and the constraints on T are given by

$$T = \begin{bmatrix} T_{11} & 0 & T_{13} & 0 \\ 0 & T_{22} & 0 & T_{24} \\ T_{31} & 0 & T_{33} & 0 \\ 0 & T_{42} & 0 & T_{44} \end{bmatrix}, \tag{11.19}$$

where the T_{ij}'s are the elements to be determined and 0's represent no connections. The question to be answered is whether for a given $4 \times m$ matrix Y, it is possible to find (non-trivial) solutions for matrix T with the structure specified by (11.19) from the matrix equation (11.14) where $\mu > 1$. We will show in the following that (non-trivial) solutions for such T always exist as long as all the diagonal elements of matrix T are assumed to be non-prespecified elements (e.g., as specified in equation (11.19)) and $p < n$ (p is defined in (11.16)). One class of sparsely interconnected neural networks which satisfies the above structural condition are the *cellular neural networks,* first introduced by Chua and Yang in 1988 [22], [23]. Cellular neural networks (which are also described by equation (11.1)), require that the matrix T have a special *sparse* structure in which all the diagonal elements are required to be non-zero.

Solutions of equation (11.14) for matrix T with prespecified zero entries will *always* exist, provided that the two conditions mentioned above are satisfied. To see this, we write (11.14) as

$$T_i Y = \mu Y_i \text{ for } i = 1, \cdots, n, \tag{11.20}$$

where T_i and Y_i represent the ith row of T and Y, respectively. For the example considered in (11.19), we have, when $i = 2$,

$$[0 \ \ T_{22} \ \ 0 \ \ T_{24}] \, Y = [0 \ \ T_{22} \ \ 0 \ \ T_{24}] \begin{bmatrix} Y_1 \\ Y_2 \\ Y_3 \\ Y_4 \end{bmatrix} = \mu Y_2.$$

This equation is equivalent to

$$[T_{22} \ \ T_{24}] \begin{bmatrix} Y_2 \\ Y_4 \end{bmatrix} = \mu Y_2, \tag{11.21}$$

and solutions of (11.21) for $[T_{22} \ \ T_{24}]$ always exist since

$$\text{rank} \begin{bmatrix} Y_2 \\ Y_4 \end{bmatrix} = \text{rank} \begin{bmatrix} Y_2 \\ Y_4 \\ \mu Y_2 \end{bmatrix}. \tag{11.22}$$

Generally speaking, when T_{ii}, $i = 1, \cdots, n$, are not prespecified as zero elements, one of the rows of Y appearing on the right hand side of (11.21) will also appear on the left hand side, which implies that the rank condition (11.22) is satisfied. The condition $p < n$ is also required, since when $p = n$, one cannot guarantee to find a non-trivial solution. (It can be proved following the procedure in [59], [87], that $p = n$ will sometimes result in the trivial solution T which is the identity matrix multiplied by μ.) Solutions of (11.21) can be determined using the singular value decomposition method as was done when solving (11.14). Specifically, we perform a singular value decomposition of

$$\begin{bmatrix} Y_2 \\ Y_4 \end{bmatrix} = [U_{i1} \vdots U_{i2}] \begin{bmatrix} D_i & \vdots & 0 \\ \cdots & \vdots & \cdots \\ 0 & \vdots & 0 \end{bmatrix} \begin{bmatrix} V_{i1}^T \\ \cdots \\ V_{i2}^T \end{bmatrix}, \tag{11.23}$$

and determine

$$[T_{22} \ T_{24}] = \mu Y_2 V_{i1} D_i^{-1} U_{i1}^T + W_i U_{i2}^T, \tag{11.24}$$

where W_i is an arbitrary row vector with appropriate dimension and the subscript $i = 2$ for the example in (11.21).

B. Summary of the Synthesis Procedure with Sparsity Constraints (non-symmetric interconnection matrix)

In the next subsection (Section 11.3C), we develop a synthesis procedure for associative memories which results in *sparse and symmetric* interconnection matrices T for system (11.1). To accomplish this, we will make use of the synthesis procedure summarized in the following (cf. Section 10.2 for notation).

Sparse Design Problem: Given an $n \times n$ index matrix $S = [S_{ij}]$ with $S_{ii} \neq 0$ for $i = 1, \cdots, n$, and m vectors $\alpha^1, \cdots, \alpha^m$ in B^n, choose $\{A, T, I\}$ with $T = T|S$ in such a manner that $\alpha^1, \cdots, \alpha^m$ are memory vectors of system (11.1). ∎

Summary of the Sparse Design Procedure:

1) Choose A as the identity matrix.

2) Choose $\mu > 1$ and $\beta^1, \cdots, \beta^m$, such that $\beta^i = \mu \alpha^i$.

3) T is solved from (11.20) or (11.21) as outlined in Section 11.3A.

4) $I = [I_1, \cdots, I_n]^T$ is computed by

$$I = \beta^m - T\alpha^m.$$

Then, $\alpha^1, \cdots, \alpha^m$ will be stored as memory vectors for system (11.1) with A, T, and I determined above. The states β^i corresponding to α^i, $i = 1, \cdots, m$, will be asymptotically stable equilibrium points of the synthesized system. ∎

Remark 11.3 From Theorem 11.1, we see that our sparse design procedure guarantees that $\alpha^1, \cdots, \alpha^m$ are also memory vectors of system (11.2) provided that

$$\|A^{-1}\Delta A\|_\infty + \|A^{-1}\Delta T\|_\infty + \|A^{-1}\Delta I\|_\infty = \|\Delta A\|_\infty + \|\Delta T\|_\infty + \|\Delta I\|_\infty < \mu - 1. \qquad (11.25)$$

The above enables us to specify *an upper bound* for the parameter inaccuracies encountered in the implementation of a given design for storing a set of desired bipolar patterns in system (11.1). This bound is chosen by the designer during the initial phase of the design procedure. This type of flexibility does not appear to have been achieved in existing synthesis procedures (e.g., [20], [30], [34], [44], [57], [58], [59], [83], [84], [87], [95], [96], [97], [121], [122]). Specifically, the synthesis procedure advocated above incorporates two features which are very important in the VLSI implementation of artificial neural networks:

. (*i*) it allows the VLSI designer to choose a suitable interconnecting structure for the neural network; and

(*ii*) it takes into account inaccuracies which arise in the realization of a neural network by hardware. ∎

In solving (11.21) for the rows of matrix T (using singular value decomposition), we encounter the unspecified row vector W_i in (11.24). In the specific examples which we consider in Section 11.4, we will choose

$$W_i = -\tau \times O_{m_i} \times U_{i2} \qquad (11.26)$$

in the Sparse Design Procedure, where $O_{m_i} = [1, \cdots, 1] \in R^{1 \times m_i}$, $m_i = \sum_{j=1}^n S_{ij}$ is the number of nonzero elements in the ith row of matrix S, and U_{i2} is obtained similarly as in (11.23).

C. Synthesis Procedure for Neural Network (11.1) with Sparsity and Symmetry Constraints on the Interconnection Matrix

For the A, T, and I determined by the Sparse Design Procedure with $\mu > 1$, let us choose

$$\Delta T = (T^T - T)/2. \qquad (11.27)$$

Then,

$$T_s \overset{\Delta}{=} T + \Delta T = (T + T^T)/2$$

is a symmetric matrix. From Theorem 11.1 (see Remark 11.2), we note that if

$$\|A^{-1}\Delta T\|_\infty = \|T^T - T\|_\infty/2 < \mu - 1, \tag{11.28}$$

the neural network (11.12) will also store all the desired patterns as memories, with a symmetric connection matrix $T + \Delta T = T_s$.

The above observation gives rise to the possibility of designing a neural network (11.1) with *prespecified interconnection structure* and with *a symmetric interconnection matrix*. (Note that in this case, we require that $S = S^T$.) Such capability is of great interest since neural network (11.1) will be *globally stable* when T is symmetric (cf. [22], [59], [65], [87], or Remark 9.4 on page 118). (Global stability means that for every initial state, the network will converge to some asymptotically stable equilibrium point and periodic and chaotic solutions do not exist.) It appears that (11.28) might be satisfied by choosing μ sufficiently large. However, from (11.24) it is seen that large μ will usually result in large absolute values of the components of T which in turn may result in a large $\|T^T - T\|_\infty$. Therefore, it is not always possible for (11.28) to be satisfied by choosing μ large. From (11.28), we see that if our original synthesized matrix T is sufficiently close to its symmetric part $(T + T^T)/2$, or equivalently, if $\|T^T - T\|_\infty$ is sufficiently small, then (11.28) is satisfied and we are able to design a neural network of form (11.1) with the following properties: (*i*) the network stores $\alpha^1, \cdots, \alpha^m$ as memory vectors; (*ii*) the network has a predetermined (full or sparse) interconnecting structure; and (*iii*) the connection matrix T of the network is symmetric.

We will utilize the following strategy to determine a matrix T which is as close to its symmetric part as possible.

Strategy: Determine the first row of T in the usual manner, *i.e.*, choose an arbitrary row vector W_1 in the computation of the first row of T (cf. (11.24)). When computing the second row, try to find a row vector W_2 such that the first component of the second row (*i.e.*, T_{21}) is equal to, or close to, the second component of the first row (*i.e.*, T_{12}). In the computations of the remaining rows of matrix T, apply a similar procedure. ∎

The synthesis procedure summarized in the previous subsection will usually result in a nonsymmetric coefficient matrix T, even when the strategy described in the preceding paragraph is used. In the following, we develop an *iterative algorithm* (design procedure) which in most cases will result in a neural network (11.1) with *symmetric* and *sparse* interconnection. In doing so, we apply Lemma 11.1 and Theorem 11.1 (Remark 11.2) *iteratively*. Let ΔT be defined as in (11.27). For the given μ (from the Sparse Design Procedure), suppose that $\|\Delta T\|_\infty \geq \mu - 1$. We can find a λ, $0 < \lambda < 1$, such that $\lambda\|\Delta T\|_\infty < \mu - 1$, and we let $T_1 = T + \lambda\Delta T$. We use this T_1 as the *new* connection matrix for our neural network (11.1). According to Lemma 11.1 and Remark 11.2 (Theorem 11.1), we see that $\alpha^1, \cdots, \alpha^m$ are still memory vectors of system (11.1) with coefficient matrix T_1, and we can compute the corresponding asymptotically stable equilibrium points as $\overline{\beta}^l = A^{-1}(T_1\alpha^l + I)$ for $l = 1, \cdots, m$. Clearly $\overline{\beta}^l \in E(\alpha^l)$. Using Theorem 11.1 (Remark 11.2), we can determine the upper bound ν for the permissible perturbation ΔT as in (11.3) and (11.4), where we use $\overline{\beta}^l$ instead of β^l. We *repeat* the above procedure, until we determine a symmetric coefficient matrix T or until we arrive at $\nu \leq 1 + \eta$ (where η is a small positive number, e.g., $\eta = 0.01$).

Because of its importance and for the sake of completeness, we summarize in the following our symmetric design procedure.

Symmetric Design Problem: Suppose we are given an index matrix $S = S^T = [S_{ij}] \in R^{n \times n}$ with $S_{ii} \neq 0$ for $i = 1, \cdots, n$, and m vectors $\alpha^1, \cdots, \alpha^m \in B^n$. Choose $\{A, T, I\}$ with $T = T|S$ and $T = T^T$ in such a manner that $\alpha^1, \cdots, \alpha^m$ are memory vectors of neural network (11.1). ∎

Symmetric Design Procedure:

1) According to the Sparse Design Procedure summerized in the previous subsection and taking into account the Strategy given above, we first choose A as the identity matrix and we determine T and I for neural network (11.1) with a specified $\mu > 1 + \eta$ (e.g., $\mu = 10$, $\eta = 0.001$).

2) If $T = T^T$ or $\mu \leq 1 + \eta$, stop. Otherwise go to step 3.

3) Compute

$$\Delta T = \frac{(T^T - T)}{2}.$$

If $\|\Delta T\|_\infty < \mu - 1$, choose $\lambda = 1$. Otherwise, choose

$$\lambda = \frac{\mu - 1}{\|\Delta T\|_\infty} - \varepsilon$$

where ε is a small positive number (e.g., $\varepsilon = 0.01$). Compute

$$T_1 = T + \lambda\Delta T.$$

4) Compute

$$\overline{\beta}^l = A^{-1}(T_1\alpha^l + I) \text{ for } l = 1, \cdots, m,$$

and compute

$$\nu = \min_{1 \le l \le m} \{\delta(\overline{\beta}^l)\} > 1.$$

5) Replacing μ by ν and replacing T by T_1, go to step 2.

If we end up with $T = T^T$, we have found a solution for our symmetric design problem. If we end up with $\mu \le 1 + \eta$ and $T \ne T^T$, our design procedure is not successful in solving a symmetric T for the given problem. ∎

The above design procedure yields a set of parameters $\{A, T, I\}$. For VLSI implementations, these parameters have to be appropriately *scaled*. The theoretical basis for doing this is provided by Corollary 9.3 on page 119.

11.4 Examples

To demonstrate the applicability and versatility of the analysis and synthesis procedures presented in this chapter, we consider two specific examples.

Example 11.1 We wish to design a neural network with 12 neurons ($n = 12$) and with the objective of storing the four patterns shown in Figure 11.2 as memories. As indicated in this figure, twelve boxes are used to represent each pattern (in R^{12}), with each box corresponding to a vector component which is allowed to assume values from -1 to 1. For purpose of visualization, -1 will represent white, 1 will represent black, and the intermediate values will correspond to appropriate grey levels, as shown in Figure 10.2 (cf. Example 10.3). The four desired patterns given in Figure 11.2 correspond to the following four bipolar vectors:

$$\alpha^1 = [1, \ 1, \ 1, \ 1, \ -1, \ -1, \ 1, \ -1, \ -1, \ 1, \ 1, \ 1]^T,$$

$$\alpha^2 = [1, \ -1, \ 1, \ 1, \ -1, \ 1, \ 1, \ 1, \ 1, \ 1, \ -1, \ 1]^T,$$

$$\alpha^3 = [-1, \ 1, \ -1, \ -1, \ 1, \ -1, \ -1, \ 1, \ -1, \ -1, \ 1, \ -1]^T,$$

and

$$\alpha^4 = [1, \ 1, \ 1, \ 1, \ -1, \ 1, \ 1, \ 1, \ 1, \ 1, \ -1, \ -1]^T.$$

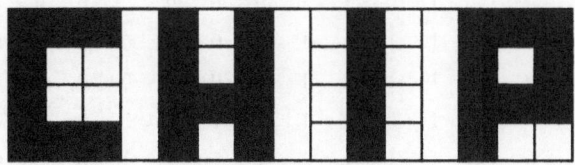

Figure 11.2: The four desired memory patterns used in Example 11.1

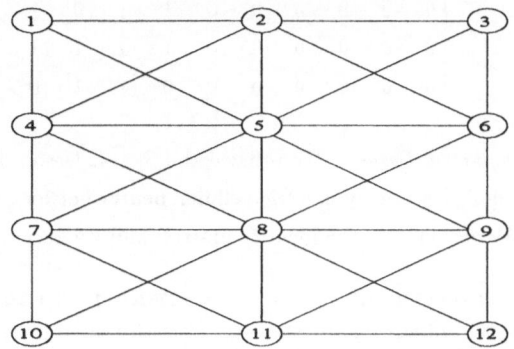

Figure 11.3: Interconnecting structure of a cellular neural network

In all cases, we seek to design a cellular neural network with the configuration given in Figure 11.3 (for details concerning cellular neural networks, see [22] or Section 10.3). The index matrix for this interconnecting structure is as follows, where "0" represents no connection and "1" represents a connection,

$$S = S^T = \begin{bmatrix} 1 & 1 & 0 & 1 & 1 & 0 & 0 & 0 & 0 & 0 & 0 & 0 \\ 1 & 1 & 1 & 1 & 1 & 1 & 0 & 0 & 0 & 0 & 0 & 0 \\ 0 & 1 & 1 & 0 & 1 & 1 & 0 & 0 & 0 & 0 & 0 & 0 \\ 1 & 1 & 0 & 1 & 1 & 0 & 1 & 1 & 0 & 0 & 0 & 0 \\ 1 & 1 & 1 & 1 & 1 & 1 & 1 & 1 & 1 & 0 & 0 & 0 \\ 0 & 1 & 1 & 0 & 1 & 1 & 0 & 1 & 1 & 0 & 0 & 0 \\ 0 & 0 & 0 & 1 & 1 & 0 & 1 & 1 & 0 & 1 & 1 & 0 \\ 0 & 0 & 0 & 1 & 1 & 1 & 1 & 1 & 1 & 1 & 1 & 1 \\ 0 & 0 & 0 & 0 & 1 & 1 & 0 & 1 & 1 & 0 & 1 & 1 \\ 0 & 0 & 0 & 0 & 0 & 0 & 1 & 1 & 0 & 1 & 1 & 0 \\ 0 & 0 & 0 & 0 & 0 & 0 & 1 & 1 & 1 & 1 & 1 & 1 \\ 0 & 0 & 0 & 0 & 0 & 0 & 0 & 1 & 1 & 0 & 1 & 1 \end{bmatrix}. \tag{11.29}$$

Case I: Nonsymmetric Design. We utilized the Sparse Design Procedure summarized in Section 11.3 to design a non-symmetric cellular neural network with the index matrix given in (11.29). We obtain A as the identity matrix, and we obtain

$$T = \begin{bmatrix} 3.3333e-01 & -1.0991e-15 & 0 & 3.3333e-01 & -1.4333e+01 & 0 \\ -3.5000e+00 & 1.5000e+01 & -3.5000e+00 & -3.5000e+00 & -1.0500e+01 & 0 \\ 0 & 0 & 5.0000e-01 & 0 & -1.4500e+01 & 1.7764e-15 \\ 2.5000e-01 & 0 & 0 & 2.5000e-01 & -1.4250e+01 & 0 \\ -5.9412e+00 & -8.3438e-16 & -5.9412e+00 & -5.9412e+00 & -8.0588e+00 & 3.5294e-01 \\ 0 & 3.1371e-15 & -5.1250e+00 & 0 & -8.8750e+00 & 5.6250e+00 \\ 0 & 0 & 0 & -2.1111e+00 & -1.1889e+01 & 0 \\ 0 & 0 & 0 & -7.1579e+00 & -6.8421e+00 & 2.1053e-01 \\ 0 & 0 & 0 & 0 & -3.1429e+00 & -7.1429e-01 \\ 0 & 0 & 0 & 0 & 0 & 0 \\ 0 & 0 & 0 & 0 & 0 & 0 \\ 0 & 0 & 0 & 0 & 0 & 0 \end{bmatrix}$$

$$\begin{bmatrix}
0 & 0 & 0 & 0 & 0 & 0 \\
0 & 0 & 0 & 0 & 0 & 0 \\
0 & 0 & 0 & 0 & 0 & 0 \\
2.5000e-01 & 0 & 0 & 0 & 0 & 0 \\
-5.9412e+00 & -7.0588e-01 & 3.5294e-01 & 0 & 0 & 0 \\
0 & 3.7500e+00 & 5.6250e+00 & 0 & 0 & 0 \\
-2.1111e+00 & -9.4444e+00 & 0 & -2.1111e+00 & -9.4444e+00 & 0 \\
-7.1579e+00 & 3.6842e-01 & 2.1053e-01 & -7.1579e+00 & -1.4211e+01 & -3.8186e-16 \\
0 & 3.1429e+00 & -7.1429e-01 & 0 & -1.3286e+01 & 5.4089e-16 \\
1.8000e+00 & -1.1400e+01 & 0 & 1.8000e+00 & -1.1400e+01 & 0 \\
-5.3750e+00 & -1.0750e+01 & -9.1250e+00 & -5.3750e+00 & -4.8750e+00 & -4.3001e-16 \\
0 & 2.1089e-16 & -7.0000e+00 & 0 & -7.0000e+00 & 1.5000e+01
\end{bmatrix},$$

and

$$I = [0,\ 1.0658e-14,\ -1.7764e-15,\ -7.1054e-15,\ 7.0588e-01,\ -3.7500e+00,\ 9.4444e+00$$

$$1.4632e+01,\ -3.1429e+00,\ 1.1400e+01,\ 1.0750e+01,\ 8.8818e-15]^T.$$

In the above computations, we chose $\mu = 15$ and W_i as in (11.26) with $\tau = 7$ for the Sparse Design Procedure. From Theorem 11.1 (Remark 11.2), we see that the upper bound for the admissible perturbation $\|\Delta T\|_\infty$ is $\mu - 1 = 14$. (For simplicity, in all of our examples, we considered $\Delta A = 0$ and $\Delta I = 0$. For the case when they are not zero, we can make similar conclusions and give similar examples.)

The performance of this network is illustrated by means of a typical simulation run of equation (11.1), shown in Figure 11.4. In this figure, the desired memory pattern is depicted in the lower right corner. The initial state, shown in the upper left corner, is generated by adding to the desired pattern zero-mean Gaussian noise with a standard deviation SD=1. The iteration of the simulation evolves from left to right in each row and from the top row to the bottom row. The desired pattern is recovered in 12 steps with a step size $h = 0.06$ in the simulation of equation (11.1).

Case II: Nonsymmetric T with Perturbations. We generated randomly a matrix $\Delta T = \Delta T | S$ as

Figure 11.4: A typical evolution of pattern No. 1 of Figure 11.2

$$\Delta T = \begin{bmatrix} -1.0794 & -1.1747 & 0 & 0.8750 & 0.3025 & 0 & 0 \\ -1.1603 & 0.6712 & -0.8235 & -1.6491 & -0.1272 & 1.1939 & 0 \\ 0 & 0.3633 & -0.0644 & 0 & 1.7388 & 1.1622 & 0 \\ 0.1312 & -1.6710 & 0 & 1.8394 & -0.1047 & 0 & -1.2227 \\ 1.6731 & -0.9829 & -0.9946 & -1.4540 & 0.9143 & 1.8469 & -1.4393 \\ 0 & 0.0208 & 0.3095 & 0 & -1.8896 & 1.6014 & 0 \\ 0 & 0 & 0 & 1.6675 & 1.0976 & 0 & -0.8437 \\ 0 & 0 & 0 & 1.0311 & -0.9883 & -0.0074 & -0.1775 \\ 0 & 0 & 0 & 0 & 1.0265 & 0.0255 & 0 \\ 0 & 0 & 0 & 0 & 0 & 0 & -1.3197 \\ 0 & 0 & 0 & 0 & 0 & 0 & -0.9107 \\ 0 & 0 & 0 & 0 & 0 & 0 & 0 \end{bmatrix}$$

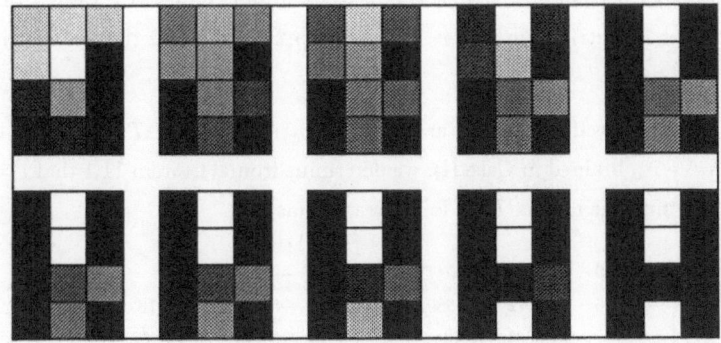

Figure 11.5: A typical evolution of pattern No. 2 of Figure 11.2

$$\begin{bmatrix} 0 & 0 & 0 & 0 & 0 \\ 0 & 0 & 0 & 0 & 0 \\ 0 & 0 & 0 & 0 & 0 \\ 0.1839 & 0 & 0 & 0 & 0 \\ -0.5019 & -1.5160 & 0 & 0 & 0 \\ 1.3049 & 1.1495 & 0 & 0 & 0 \\ -1.2986 & 0 & -1.2319 & -0.9470 & 0 \\ -1.6428 & 1.6352 & -0.5880 & 0.1074 & -1.9514 \\ -1.7075 & -0.8249 & 0 & 1.1790 & -0.7365 \\ -1.2635 & 0 & 1.1235 & -0.4117 & 0 \\ 0.0318 & -0.7197 & -0.8434 & 0.3578 & -1.8620 \\ -1.2534 & 0.5991 & 0 & 1.6019 & 0.8208 \end{bmatrix}$$

which satisfies the condition that $\|\Delta T\|_\infty < \mu - 1$. We use $T_2 \triangleq T + \Delta T$ in system (11.12).

Since $\|\Delta T\|_\infty < \mu - 1$, we see from Theorem 11.1 (Remark 11.2) that $\alpha^1, \cdots, \alpha^4$ are also memories for system (11.12). A typical simulation run of equation (11.12) with ΔT given above is depicted in Figure 11.5. In this figure, the noisy pattern is generated by adding to the desired pattern uniformly distributed noise defined on $[-1, 1]$. Convergence occurs in 9 steps with $h = 0.06$.

Case III: Symmetric Design. Using the Symmetric Design Procedure outlined in Section 11.3C, we can easily determine a symmetric matrix T for the present design. We begin with the T matrix obtained in Case I. (Note that we do not consider the Strategy given in Section 11.3C.) Choosing $\varepsilon = 0.01$ and $\eta = 0.001$ in our Symmetric Design Procedure, we find a symmetric matrix T in four iterations (step 2 to step 5 of the Symmetric Design Procedure).

With ε and η as specified above, starting with matrix $T_2 = T + \Delta T$ (where T is obtained in Case I and ΔT is obtained in Case II), we determine from Theorem 11.1 that $\nu = 9.5512$, and we find a symmetric matrix T_3 in four iterations as

$$T_3 = \begin{bmatrix}
-0.7460 & -2.9175 & 0 & 0.7948 & -9.1495 & 0 & 0 \\
-2.9175 & 15.6712 & -1.9801 & -3.4101 & -5.8051 & 0.6073 & 0 \\
0 & -1.9801 & 0.4356 & 0 & -9.8485 & -1.8267 & 0 \\
0.7948 & -3.4101 & 0 & 2.0894 & -10.8749 & 0 & -0.7082 \\
-9.1495 & -5.8051 & -9.8485 & -10.8749 & -7.1446 & -4.2824 & -9.0859 \\
0 & 0.6073 & -1.8267 & 0 & -4.2824 & 7.2264 & 0 \\
0 & 0 & 0 & -0.7082 & -9.0859 & 0 & -2.9548 \\
0 & 0 & 0 & -2.9714 & -4.5191 & 2.6290 & -9.0392 \\
0 & 0 & 0 & 0 & -1.6397 & 3.0429 & 0 \\
0 & 0 & 0 & 0 & 0 & 0 & -1.4314 \\
0 & 0 & 0 & 0 & 0 & 0 & -8.3386 \\
0 & 0 & 0 & 0 & 0 & 0 & 0
\end{bmatrix}$$

$$\begin{bmatrix}
0 & 0 & 0 & 0 & 0 \\
0 & 0 & 0 & 0 & 0 \\
0 & 0 & 0 & 0 & 0 \\
-2.9714 & 0 & 0 & 0 & 0 \\
-4.5191 & -1.6397 & 0 & 0 & 0 \\
2.6290 & 3.0429 & 0 & 0 & 0 \\
-9.0392 & 0 & -1.4314 & -8.3386 & 0 \\
-1.2743 & 1.6406 & -10.2047 & -12.4107 & -1.6024 \\
1.6406 & -1.5392 & 0 & -10.9757 & -3.5687 \\
-10.2047 & 0 & 2.9235 & -9.0150 & 0 \\
-12.4107 & -10.9757 & -9.0150 & -4.5172 & -3.6301 \\
-1.6024 & -3.5687 & 0 & -3.6301 & 15.8208
\end{bmatrix}.$$

It can be verified by Lemma 11.1 that $\alpha^1, \alpha^2, \alpha^3$, and α^4 are also memories for system (11.1) with the symmetric matrix $T = T_3$ given above, and that the corresponding asymptotically stable equilibrium points are given by

$$\beta^1 = [6.2807, \ 12.5613, \ 10.1306, \ 12.6123, \ -26.4722, \ -13.5853, \ 14.1366, \ -20.0730$$

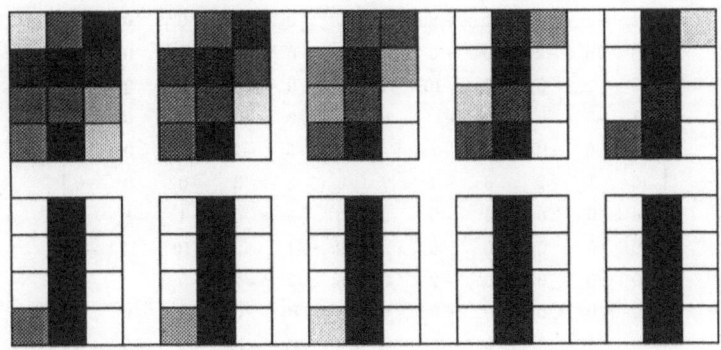

Figure 11.6: A typical evolution of pattern No. 3 of Figure 11.2

$$-19.1918, \ 14.0818, \ 8.6355, \ 17.3618]^{T}$$

$$\beta^{2} = [12.1157, \ -17.5664, \ 10.4374, \ 13.4895, \ -35.7444, \ 10.9967, \ 12.7353, \ 10.7388$$

$$9.0481, \ 11.7025, \ -29.1028, \ 14.2798]^{T}$$

$$\beta^{3} = [-12.1157, \ 17.5664, \ -10.4374, \ -19.4324, \ 28.1179, \ -13.2387, \ -11.9249, \ 15.9757$$

$$-12.0527, \ -9.3119, \ 25.7815, \ -17.4846]^{T}$$

and

$$\beta^{4} = [6.2807, \ 13.7760, \ 6.4773, \ 6.6694, \ -47.3545, \ 12.2114, \ 12.7353, \ 13.9436$$

$$16.1855, \ 11.7025, \ -21.8427, \ -17.3618]^{T}$$

From Theorem 11.1 (Remark 11.2), we can verify that the allowable upper bound for the perturbation for system (11.1) with the above symmetric matrix T_3 is given by $\|\Delta T\|_{\infty} < 6.2807 - 1 = 5.2807$.

A typical simulation run for system (11.1) with T_3 given above is shown in Figure 11.6. In this case, the noisy pattern is generated by adding Gaussian noise $N(0, 1)$ to the desired pattern. Convergence occurs in 8 steps with $h = 0.06$.

Case IV: Rounded Matrix T. We round every component of matrix T_3 obtained in Case III to its closest integer and obtain the matrix T_4 given by

$$T_4 = \begin{bmatrix} -1 & -3 & 0 & 1 & -9 & 0 & 0 & 0 & 0 & 0 & 0 & 0 \\ -3 & 16 & -2 & -3 & -6 & 1 & 0 & 0 & 0 & 0 & 0 & 0 \\ 0 & -2 & 0 & 0 & -10 & -2 & 0 & 0 & 0 & 0 & 0 & 0 \\ 1 & -3 & 0 & 2 & -11 & 0 & -1 & -3 & 0 & 0 & 0 & 0 \\ -9 & -6 & -10 & -11 & -7 & -4 & -9 & -5 & -2 & 0 & 0 & 0 \\ 0 & 1 & -2 & 0 & -4 & 7 & 0 & 3 & 3 & 0 & 0 & 0 \\ 0 & 0 & 0 & -1 & -9 & 0 & -3 & -9 & 0 & -1 & -8 & 0 \\ 0 & 0 & 0 & -3 & -5 & 3 & -9 & -1 & 2 & -10 & -12 & -2 \\ 0 & 0 & 0 & 0 & -2 & 3 & 0 & 2 & -2 & 0 & -11 & -4 \\ 0 & 0 & 0 & 0 & 0 & 0 & -1 & -10 & 0 & 3 & -9 & 0 \\ 0 & 0 & 0 & 0 & 0 & 0 & -8 & -12 & -11 & -9 & -5 & -4 \\ 0 & 0 & 0 & 0 & 0 & 0 & 0 & -2 & -4 & 0 & -4 & 16 \end{bmatrix}. \tag{11.30}$$

Using Theorem 11.1 (Remark 11.2), we can see that with the matrix T_4 given above, which consists of integers and which is also symmetric, the desired patterns α^l, $l = 1, 2, 3, 4$, are also memories of system (11.1) with T_4 given in (11.30), since the perturbation which we used to obtain the above $T_4 = T_3 + \Delta T$ satisfies $\|\Delta T\|_\infty < 5.2807$. We can determine the permissible upper bound for the perturbation ΔT to the matrix T_4 in (11.30) as $\|\Delta T\|_\infty < 5$ (by Theorem 11.1).

A typical simulation run for the present case is depicted in Figure 11.7. In this figure, the noisy pattern is generated by adding Gaussian noise $N(0.1, 1)$ to the desired pattern. Convergence occurs in 14 steps with $h = 0.06$. ∎

Example 11.2 In order to test our Symmetric Design Procedure and to see how typical the results of Case III in Example 11.1 are, we repeated these examples 200 times using different sets of desired patterns to be stored as memory vectors. Each set contains $m = 4$ vectors in B^{12} which were generated randomly. For each given set of vectors, we synthesized system (11.1) using the Symmetric Design Procedure.

In these 200 tests, we chose $\mu = 10$ and W_i as in (11.26) with $\tau = 1$ in the Symmetric Design Procedure. There are only 11 tests out of 200 in which we did not succeed in finding a symmetric matrix T for the generated desired patterns and using the above specifications (for μ and W_i). Furthermore, for these 11 failed tests, when we increased μ from 10 to 15, we were able to determine symmetric matrices T again. ∎

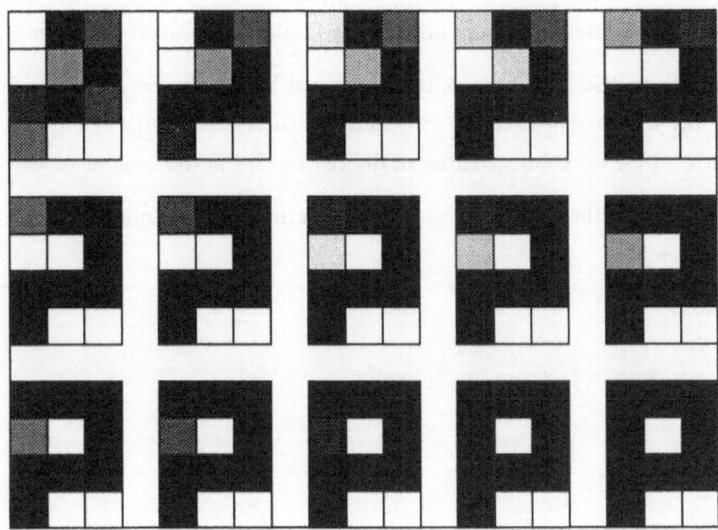

Figure 11.7: A typical evolution of pattern No. 4 of Figure 11.2

Results in Example 11.2 suggest that our Symmetric Design Procedure will frequently succeed in determining a symmetric matrix T for the design problem considered herein. It also suggests that there might be a tradeoff between μ and τ (when we choose W_i as given in (11.26)). Usually, choosing a larger μ and a smaller τ makes it easier to find a symmetric T. The disadvantages resulting from such choices of μ and τ are as follows. A large μ will usually result in a matrix T having components with large absolute values. In VLSI implementations of neural networks, we usually want to avoid large values for the parameters (since they correspond to amplifications). Also, experimental results showed that smaller τ will result in synthesized networks with more spurious memories.

11.5   Concluding Remarks

The results of the present chapter complement our results in Chapter 10, [63], [65], and [70] on sparsely interconnected neural networks. They also constitute an extension of our results in [71] to sparsely interconnected neural networks:

(i) We provide upper bounds for the perturbations of parameters under which desired memories stored in a neural network (11.1) are preserved (Theorem 11.1). This type of information is of great practical interest during the implementation process of such networks.

(ii) The Symmetric Design Procedure presented herein enables us to design artificial neural networks with prespecified interconnecting structure and with symmetric interconnection matrix which store any given set of desired bipolar patterns as memories.

We demonstrated the applicability and the versatility of our results by means of several specific examples.

CLOSING REMARKS

Several fundamental issues concerning the qualitative behavior of a class of nonlinear systems–systems with saturation type nonlinearities–have been addressed in this monograph. We investigated extensively certain important problems in control systems, signal processing, and neural networks. All of our research topics are concerned with classes of systems with saturation type nonlinearities, and we believe that good progress was made on each topic. This work makes fundamental contributions to the *Qualitative Theory of Dynamical Systems with Saturation Nonlinearities.* In particular, our work makes contributions to the following:

- Asymptotic stability analysis of continuous-time and discrete-time dynamical systems with saturation nonlinearities.

- Stabilizability and null controllability of systems with state saturation and control constraints.

- Asymptotic stability analysis of fixed-point digital filters using generalized overflow nonlinearities.

- Asymptotic stability analysis of multidimensional state-space digital filters endowed with overflow nonlinearities.

- Analysis and synthesis of a class of neural networks with piecewise linear saturation activation functions.

- Synthesis procedures for neural networks with prespecified connectivity constraints.

- Robustness analysis and design of neural networks with sparse interconnections.

BIBLIOGRAPHY

[1] T. T. Aboulnasr, M. M. Fahmy, "Finite-word-length effects in two-dimensional digital systems," in *Multidimensional Systems: Techniques and Applications*, Edited by S. G. Tzafestas, New York: Marcel Dekker, 1986

[2] P. Agathoklis, E. I. Jury, M. Mansour, "Criteria for the absence of limit cycles in two-dimensional discrete systems," *IEEE Transactions on Acoustics, Speech, and Signal Processing*, vol. ASSP-32, pp. 432–434, Apr. 1984

[3] S. V. B. Aiyer, M. Niranjan, F. Fallside, "A theoretical investigation into the performance of the Hopfield model," *IEEE Transactions on Neural Networks*, vol. 1, pp. 204–215, June 1990

[4] K. J. Åström, B. W. Wittenmark, *Computer Controlled Systems: Theory and Design*, Englewood Cliffs, NJ: Prentice-Hall, 1984

[5] G. Avitabile, M. Forti, S. Manetti, M. Marini, "On a class of nonsymmetrical neural networks with application to ADC," *IEEE Transactions on Circuits and Systems*, vol. 38, pp. 202–209, Feb. 1991

[6] C. W. Barnes, A. T. Fam, "Minimum norm recursive digital filters that are free of overflow limit cycles," *IEEE Transactions on Circuits and Systems*, vol. CAS-24, pp. 569–574, Oct. 1977

[7] P. H. Bauer, E. I. Jury, "Stability analysis of multidimensional (m-D) direct realization digital filters under the influence of nonlinearities," *IEEE Transactions on Acoustics, Speech, and Signal Processing*, vol. 36, pp. 1770–1780, Nov. 1988

[8] P. H. Bauer, E. I. Jury, "A stability analysis of two-dimensional nonlinear digital state-space filters," *IEEE Transactions on Acoustics, Speech, and Signal Processing*, vol. 38, pp. 1578–1586, Sept. 1990

[9] A. Benzaouia, C. Burgat, "Regulator problem for linear discrete-time systems with non-symmetrical constrained control," *Int. J. Control*, vol. 48, pp. 2441–2451, Dec. 1988

[10] F. Blanchini, "Feedback control for linear time-invariant systems with state and control bounds in the presence of disturbances," *IEEE Transactions on Automatic Control*, vol. 35, pp. 1231–1234, Nov. 1990

[11] A. G. Bolton, "A two's complement overflow limit cycle free digital filter structure," *IEEE Transactions on Circuits and Systems*, vol. CAS-31, pp. 1045–1046, Dec. 1984

[12] T. Bose, Mei-Qin Chen, "Overflow oscillations in state-space digital filters," *IEEE Transactions on Circuits and Systems*, vol. 38, pp. 807–810, July 1991

[13] J. Bruck, "On the convergence properties of the Hopfield model," *Proceedings of the IEEE*, vol. 78, pp. 1579–1585, Oct. 1990

[14] G. A. Carpenter, M. A. Cohen, S. Grossberg, "Computing with neural networks," *Science*, vol. 235, pp. 1226–1227, 1987

[15] Bor-Sen Chen, Sin-Syung Wang, "The design of feedback controller with nonlinear saturating actuator: Time domain approach," *Proceedings of the 25th IEEE Conference on Decision and Control*, Athens, Greece, pp. 2048–2053, Dec. 1986

[16] Bor-Sen Chen, Ching-Chang Wong, "Robust stabilization of multivariable systems via constrained control: Time domain approach," *Proceedings of the 26th IEEE Conference on Decision and Control*, Los Angeles, CA, pp. 1292–1297, Dec. 1987

[17] Jyh-Horng Chou, "Stabilization of linear discrete-time systems with actuator saturation," *Systems & Control Letters*, vol. 17, pp. 141–144, Aug. 1991

[18] L. O. Chua, T. Roska, "Stability of a class of nonreciprocal cellular neural networks," *IEEE Transactions on Circuits and Systems*, vol. 37, pp. 1520–1527, Dec. 1990

[19] L. O. Chua, T. Roska, "The CNN paradigm," *IEEE Transactions on Circuits and Systems-I: Fundamental Theory and Applications*, vol. 40, pp. 147–156, Mar. 1993

[20] L. O. Chua, P. Thiran, "An analytic method for designing simple cellular neural networks," *IEEE Transactions on Circuits and Systems*, vol. 38, pp. 1332–1341, Nov. 1991

[21] L. O. Chua, C. W. Wu, "On the universe of stable cellular neural networks," *Int. J. Circuit Theory and Applications*, vol. 20, pp. 497–572, 1992

[22] L. O. Chua, L. Yang, "Cellular neural networks: Theory," *IEEE Transactions on Circuits and Systems*, vol. 35, pp. 1257–1272, Oct. 1988

[23] L. O. Chua, L. Yang, "Cellular neural networks: Applications," *IEEE Transactions on Circuits and Systems*, vol. 35, pp. 1273–1290, Oct. 1988

[24] T. A. C. Claasen, L. O. G. Kristiansson, "Necessary and sufficient conditions for the absence of overflow phenomena in a second-order recursive digital filter," *IEEE Transactions on Acoustics, Speech, and Signal Processing*, vol. ASSP-23, pp. 509–515, Dec. 1975

[25] T. A. C. Claasen, W. F. G. Mecklenbräuker, J. B. H. Peek, "Frequency domain criteria for the absence of zero-input limit cycles in nonlinear discrete-time systems, with applications to digital filters," *IEEE Transactions on Circuits and Systems*, vol. CAS-22, pp. 232–239, March 1975

[26] T. A. C. Claasen, W. F. G. Mecklenbräuker, J. B. H. Peek, "Effects of quantization and overflow in recursive digital filters," *IEEE Transactions on Acoustics, Speech, and Signal Processing*, vol. ASSP-24, pp. 517–529, Dec. 1976

[27] M. A. Cohen, S. Grossberg, "Absolute stability of global pattern formation and parallel memory storage by competitive neural networks," *IEEE Transactions on Systems, Man, and Cybernetics*, vol. 13, pp. 815–826, Sept./Oct. 1983

[28] W. A. Coppel, *Stability and Asymptotic Behavior of Differential Equations*, Boston, MA: D. C. Heath, 1965

[29] M. Cottrell, "Stability and attractivity in associative memory networks," *Biol. Cybern.*, vol. 58, pp. 129–139, 1988

[30] S. R. Das, "On the synthesis of nonlinear continuous neural networks," *IEEE Transactions on Systems, Man, and Cybernetics*, vol. 21, pp. 413–418, March/Apr. 1991

[31] P. M. Ebert, J. E. Mazo, M. G. Taylor, "Overflow oscillations in digital filters," *The Bell System Technical Journal*, vol. 48, pp. 2999–3020, Nov. 1969

[32] N. G. El-Agizi, M. M. Fahmy, "Two-dimensional digital filters with no overflow oscilla-tions," *IEEE Transactions on Acoustics, Speech, and Signal Processing*, Vol. ASSP-27, pp. 465–469, Oct. 1979

[33] K. T. Erickson, A. N. Michel, "Stability analysis of fixed-point digital filters using computer generated Lyapunov functions–Part I: Direct form and coupled form filters," *IEEE Transactions on Circuits and Systems*, vol. CAS-32, pp. 113–132, Feb. 1985

[34] J. A. Farrell, A. N. Michel, "A synthesis procedure for Hopfield's continuous-time associative memory," *IEEE Transactions on Circuits and Systems*, vol. 37, pp. 877–884, July 1990

[35] J. F. Frankena, R. Sivan, "A nonlinear optimal control law for linear systems," *Int. J. Control*, vol. 30, pp. 159–178, July 1979

[36] G. F. Franklin, J. D. Powell, *Digital Control of Dynamic Systems*, Reading, MA: Addison-Wesley, 1980

[37] A. H. Glattfelder, W. Schaufelberger, "Stability analysis of single loop control systems with saturation and antireset-windup circuits," *IEEE Transactions on Automatic Con-trol*, vol. AC-28, pp. 1074–1081, Dec. 1983

[38] R. M. Golden, "The 'brain-state-in-a-box' neural model is a gradient descent algo-rithm," *J. of Mathematical Psychology*, vol. 30, pp. 73–80, March 1986

[39] H. J. Greenberg, "Equilibria of the brain-state-in-a-box (BSB) neural model," *Neural Networks*, vol. 1, pp. 323–324, 1988

[40] S. Grossberg, "Nonlinear neural networks: Principles, mechanisms, and architectures," *Neural Networks*, vol. 1, pp. 17–61, 1988

[41] L. T. Grujić, A. N. Michel, "Exponential stability and trajectory bounds of neural networks under structure variations," *IEEE Transactions on Circuits and Systems*, vol. 38, pp. 1182–1192, Oct. 1991

[42] A. Guez, V. Protopopsecu, J. Barhen, "On the stability, storage capacity, and design of nonlinear continuous neural networks," *IEEE Transactions on Systems, Man, and Cybernetics*, vol. 18, pp. 80–87, Jan./Feb. 1988

[43] P.-O. Gutman, P. Hagander, "A new design of constrained controllers for linear systems," *IEEE Transactions on Automatic Control*, vol. AC-30, pp. 22–33, Jan. 1985

[44] J. J. Hopfield, "Neural networks and physical systems with emergent collective computational abilities," *Proc. Nat. Acad. Sci. USA*, vol. 79, pp. 2554–2558, Apr. 1982

[45] J. J. Hopfield, "Neurons with graded response have collective computational properties like those of two-state neurons," *Proc. Nat. Acad. Sci. USA*, vol. 81, pp. 3088–3092, May 1984

[46] J. J. Hopfield, D. W. Tank, " 'Neural' computation of decisions in optimization problems," *Biol. Cybern.*, vol. 52, pp. 141–152, 1985

[47] S. Hui, S. H. Żak, "Dynamical analysis of the brain-state-in-a-box (BSB) neural models," *IEEE Transactions on Neural Networks*, vol. 3, pp. 86–94, Jan. 1992

[48] R. E. Kalman, Y. C. Ho, K. S. Narendra, "Controllability of linear dynamical systems," *Contributions to Differential Equations*, vol. 11, no. 2, pp. 189–213, 1963

[49] P. Kapasouris, M. Athans, G. Stein, "Design of feedback control systems for stable plants with saturating actuators," *Proceedings of the 27th IEEE Conference on Decision and Control*, Austin, TX, pp. 469–479, Dec. 1988

[50] E. Kaszkurewicz, A. Bhaya, "Comments on 'Overflow oscillations in state-space digital filters'," *IEEE Transactions on Circuits and Systems-II: Analog and Digital Signal Processing*, vol. 39, pp. 675–676, Sept. 1992

[51] M. Kawamata, T. Higuchi, "On the absence of limit cycles in a class of state-space digital filters which contains minimum noise realizations," *IEEE Transactions on Acoustics, Speech, and Signal Processing*, vol. ASSP-32, pp. 928–930, Aug. 1984

[52] J. D. Keeler, "Basins of attraction of neural network models," *AIP Conference Proceedings 151*, Snowbird, UT, pp. 259–264, 1986

[53] R. L. Kosut, "Design of linear systems with saturating linear control and bounded states," *IEEE Transactions on Automatic Control*, vol. AC-28, pp. 121–124, Jan. 1983

[54] N. J. Krikelis, S. K. Barkas, "Design of tracking systems subject to actuator saturation and integrator wind-up," *Int. J. Control*, vol. 39, pp. 667–682, Apr. 1984

[55] E. B. Lee, L. Markus, *Foundations of Optimal Control Theory*, New York, NY: John Wiley and Sons, 1967

[56] J. Levendovszky, "A possible transformation of fully connected neural nets into partially connected networks," *Proc. 1990 IEEE Int. Workshop on Cellular Neural Networks and Their Applications,* Budapest, Hungary, pp. 55–64, Dec. 1990

[57] J.-H. Li, A. N. Michel, W. Porod, "Qualitative analysis and synthesis of a class of neural networks," *IEEE Transactions on Circuits and Systems*, vol. 35, pp. 976–986, Aug. 1988

[58] J.-H. Li, A. N. Michel, W. Porod, "Analysis and synthesis of a class of neural networks: Variable structure systems with infinite gain," *IEEE Transactions on Circuits and Systems*, vol. 36, pp. 713–731, May 1989

[59] J.-H. Li, A. N. Michel, W. Porod, "Analysis and synthesis of a class of neural networks: Linear systems operating on a closed hypercube," *IEEE Transactions on Circuits and Systems*, vol. 36, pp. 1405–1422, Nov. 1989

[60] Derong Liu, A. N. Michel, "Asymptotic stability of systems operating on a closed hypercube," *Systems & Control Letters*, vol. 19, pp. 281–285, Oct. 1992

[61] Derong Liu, A. N. Michel, "Asymptotic stability of discrete-time systems with saturation nonlinearities with applications to digital filters," *IEEE Transactions on Circuits and Systems-I: Fundamental Theory and Applications*, vol. 39, pp. 798–807, Oct. 1992

[62] Derong Liu, A. N. Michel, "Null controllability of systems with control constraints and state saturation," *Systems & Control Letters*, vol. 20, pp. 131–139, Feb. 1993

[63] Derong Liu, A. N. Michel, "Cellular neural networks for associative memories," *IEEE Transactions on Circuits and Systems-II: Analog and Digital Signal Processing*, vol. 40, pp. 119–121, Feb. 1993

[64] Derong Liu, A. N. Michel, "Stability analysis of state-space realizations for two-dimensional filters with overflow nonlinearities," *IEEE Transactions on Circuits and Systems-I: Fundamental Theory and Applications*, to appear

[65] Derong Liu, A. N. Michel, "Sparsely interconnected neural networks for associative memories with applications to cellular neural networks," *IEEE Transactions on Circuits and Systems-II: Analog and Digital Signal Processing*, vol. 41, Apr. 1994, to appear

[66] Derong Liu, A. N. Michel, "Robustness analysis and design of a class of neural networks with sparse interconnecting structure," submitted to *IEEE Transactions on Circuits and Systems*

[67] Derong Liu, A. N. Michel, "Asymptotic stability of fixed point digital filters using generalized overflow arithmetic," *Proceedings of the 30th Annual Allerton Conference on Communication, Control, and Computing*, University of Illinois at Urbana-Champaign, Urbana, IL, pp. 485–494, Sept. 1992

[68] Derong Liu, A. N. Michel, "Asymptotic stability of discrete-time systems with saturation nonlinearities," *Proceedings of the 31st IEEE Conference on Decision and Control*, Tucson, AZ, pp. 3440–3445, Dec. 1992

[69] Derong Liu, A. N. Michel, "Asymptotic stability of two-dimensional digital filters with overflow nonlinearities," *Proceedings of the 1993 International Symposium on Circuits and Systems*, Chicago, IL, pp. 575–578, May 1993

[70] Derong Liu, A. N. Michel, "Analysis and synthesis of a class of neural networks with sparse interconnection," *Proceedings of the 1993 International Symposium on Circuits and Systems*, Chicago, IL, pp. 2596–2599, May 1993

[71] Derong Liu, A. N. Michel, "Robustness analysis of a class of neural networks," *Proceedings of the 36th Midwest Symposium on Circuits and Systems*, Detroit, MI, Aug. 1993

[72] Derong Liu, A. N. Michel, "Analysis and synthesis of sparsely connected feedback neural networks," accepted by *the 32nd IEEE Conference on Decision and Control*, San Antonio, TX, Dec. 1993

[73] Derong Liu, A. N. Michel, "Robustness analysis and design of sparsely interconnected neural networks," accepted by *the 32nd IEEE Conference on Decision and Control*, San Antonio, TX, Dec. 1993

[74] M. Marcus, H. Minc, *A Survey of Matrix Theory and Matrix Inequalities*, Boston, MA: Allyn and Bacon, 1964

[75] C. M. Marcus, R. M. Westervelt, "Dynamics of iterated-map neural networks," *Physical Review A*, vol. 40, pp. 501–504, July 1989

[76] T. Matsumoto, L. O. Chua, R. Furukawa, "CNN cloning template: Hole-filler," *IEEE Transactions on Circuits and Systems*, vol. 37, pp. 635–638, May 1990

[77] T. Matsumoto, L. O. Chua, H. Suzuki, "CNN cloning template: Connected component detector," *IEEE Transactions on Circuits and Systems*, vol. 37, pp. 633–635, May 1990

[78] T. Matsumoto, L. O. Chua, H. Suzuki, "CNN cloning template: Shadow detector," *IEEE Transactions on Circuits and Systems*, vol. 37, pp. 1070–1073, Aug. 1990

[79] T. Matsumoto, L. O. Chua, T. Yokohama, "Image thinning with a cellular neural network," *IEEE Transactions on Circuits and Systems*, vol. 37, pp. 638–640, May 1990

[80] K. Matsuoka, "Stability conditions for nonlinear continuous neural networks with asymmetric connection weights," *Neural Networks*, vol. 5, pp. 495–500, 1992

[81] A. N. Michel, J. A. Farrell, "Associative memories via artificial neural networks," *IEEE Control Systems Magazine*, vol. 10, pp. 6–17, Apr. 1990

[82] A. N. Michel, J. A. Farrell, W. Porod, "Qualitative analysis of neural networks," *IEEE Transactions on Circuits and Systems*, vol. 36, pp. 229–243, Feb. 1989

[83] A. N. Michel, J. A. Farrell, H. -F. Sun, "Analysis and synthesis techniques for Hopfield type synchronous discrete time neural networks with application to associative memory," *IEEE Transactions on Circuits and Systems*, vol. 37, pp. 1356–1366, Nov. 1990

[84] A. N. Michel, D. L. Gray, "Analysis and synthesis of neural networks with lower block triangular interconnecting structure," *IEEE Transactions on Circuits and Systems*, vol. 37, pp. 1267–1283, Oct. 1990

[85] A. N. Michel, C. J. Herget, *Applied Algebra and Functional Analysis*, New York, NY: Dover Publications, Inc., 1993

[86] A. N. Michel, R. K. Miller, *Qualitative Analysis of Large Scale Dynamical Systems*, New York, NY: Academic, 1977

[87] A. N. Michel, Jie Si, Gune Yen, "Analysis and synthesis of a class of discrete-time neural networks described on hypercubes," *IEEE Transactions on Neural Networks*, vol. 2, pp. 32–46, Jan. 1991

[88] R. K. Miller, A. N. Michel, *Ordinary Differential Equations*, New York: Academic, 1982

[89] R. K. Miller, M. S. Mousa, A. N. Michel, "Quantization and overflow effects in digital implementations of linear dynamic controllers," *IEEE Transactions on Automatic Control*, vol. 33, pp. 698–704, July 1988

[90] W. L. Mills, C. T. Mullis, R. A. Roberts, "Digital filter realizations without overflow oscillations," *IEEE Transactions on Acoustics, Speech, and Signal Processing*, vol. ASSP-26, pp. 334–338, Aug. 1978

[91] D. Mitra, "Large amplitude, self-sustained oscillations in difference equations which describe digital filter sections using saturation arithmetic," *IEEE Transactions on Acoustics, Speech, and Signal Processing*, vol. ASSP-25, pp. 134–143, Apr. 1977

[92] D. Mitra, "Criteria for determining if a high-order digital filter using saturation arithmetic is free of overflow oscillations," *The Bell System Technical Journal*, vol. 56, pp. 1679–1699, Nov. 1977

[93] K. Ogata, *State Space Analysis of Control Systems*, Englewood Cliffs, NJ: Prentice-Hall, 1967

[94] L. Pandolfi, "Linear control systems: Controllability with constrained controls," *J. Optimization Theory and Applications*, vol. 19, pp. 577–585, Aug. 1976

[95] R. Perfetti, "A neural network to design neural networks," *IEEE Transactions on Circuits and Systems*, vol. 38, pp. 1099–1103, Sept. 1991

[96] L. Personnaz, I. Guyon, G. Dreyfus, "Information storage and retrieval in spin-glass like neural networks," *J. Physique Lett.*, vol. 46, pp. 359–365, Apr. 1985

[97] L. Personnaz, I. Guyon, G. Dreyfus, "Collective computational properties of neural networks: New learning mechanisms," *Physical Review A*, vol. 34, pp. 4217–4228, Nov. 1986

[98] J. H. F. Ritzerfeld, "A condition for the overflow stability of second-order digital filters that is satisfied by all scaled state-space structures using saturation," *IEEE Transactions on Circuits and Systems*, vol. 36, pp. 1049–1057, Aug. 1989

[99] R. P. Roesser, "A discrete state-space model for linear image processing," *IEEE Transactions on Automatic Control*, vol. AC-20, pp. 1–10, Feb. 1975

[100] T. Roska, L. O. Chua, "Cellular neural networks with nonlinear and delay-type template elements," *Int. J. Circuit Theory and Applications*, vol. 20, pp. 469–481, 1992

[101] F. M. A. Salam, Y. Wang, M.-R. Choi, "On the analysis of dynamic feedback neural nets," *IEEE Transactions on Circuits and Systems*, vol. 38, pp. 196–201, Feb. 1991

[102] I. W. Sandberg, "A theorem concerning limit cycles in digital filters," *Proceedings of the 7th Annual Allerton Conference on Circuit and System Theory*, University of Illinois at Urbana-Champaign, Urbana, IL, pp. 63–68, Oct. 1969

[103] I. W. Sandberg, "The zero-input response of digital filters using saturation arithmetic," *IEEE Transactions on Circuits and Systems*, vol. CAS-26, pp. 911–915, Nov. 1979

[104] M. E. Savran, Ö. Morgül, "On the associative memory design for the Hopfield neural network," *Proceedings of the 1991 IEEE International Joint Conference on Neural Networks*, Singapore, pp. 1166–1171, Nov. 1991

[105] W. E. Schmitendorf, B. R. Barmish, "Null controllability of linear systems with constrained controls," *SIAM J. Control and Optimization*, vol. 18, pp. 327–345, July 1980

[106] V. Singh, "A new realizability condition for limit cycle-free state-space digital filters employing saturation arithmetic," *IEEE Transactions on Circuits and Systems*, vol. CAS-32, pp. 1070–1071, Oct. 1985

[107] V. Singh, "Realization of two's complement overflow limit cycle free state-space digital filters: A frequency-domain viewpoint," *IEEE Transactions on Circuits and Systems*, vol. CAS-33, pp. 1042–1044, Oct. 1986

[108] V. Singh, "Elimination of overflow oscillations in fixed-point state-space digital filters using saturation arithmetic," *IEEE Transactions on Circuits and Systems*, vol. 37, pp. 814–818, June 1990

[109] E. D. Sontag, "An algebraic approach to bounded controllability of linear systems," *Int. J. Control*, vol. 39, pp. 181–188, Jan. 1984

[110] E. D. Sontag, *Mathematical Control Theory: Deterministic Finite Dimensional Systems*, New York, NY: Springer-Verlag, 1990

[111] M. Sznaier, M. J. Damborg, "Control of constrained discrete time linear systems using quantized controls," *Automatica*, vol. 25, pp. 623–628, July 1989

[112] D. W. Tank, J. J. Hopfield, "Simple 'neural' optimization networks: An A/D converter, signal decision circuit, and a linear programming circuit," *IEEE Transactions on Circuits and Systems*, vol. CAS-33, pp. 533–541, May 1986

[113] S. G. Tzafestas, A. Kanellakis, N. J. Theodorou, "Two-dimensional digital filters without overflow oscillations and instability due to finite word length," *IEEE Transactions on Signal Processing*, vol. 40, pp. 2311–2317, Sept. 1992

[114] P. P. Vaidyanathan, V. Liu, "An improved sufficient condition for absence of limit cycles in digital filters," *IEEE Transactions on Circuits and Systems*, vol. CAS-34, pp. 319–322, March 1987

[115] R. P. Van Til, W. E. Schmitendorf, "Constrained controllability of discrete-time systems," *Int. J. Control*, vol. 43, pp. 941–956, March 1986

[116] M. Vassilaki, J. C. Hennet, G. Bitsoris, "Feedback control of linear discrete-time systems under state and control constraints," *Int. J. Control*, vol. 47, pp. 1727–1735, June 1988

[117] G. R. Walsh, *An Introduction to Linear Programming*, New York, NY: John Wiley and Sons, 1985

[118] F. R. Waugh, C. M. Marcus, R. M. Westervelt, "Fixed-point attractors in analog neural computation," *Physical Review Letters*, vol. 64, pp. 1986–1989, Apr. 1990

[119] B. Widrow, "30 years of adaptive neural networks: Perceptron, madaline, and back-propagation," *Proceedings of the IEEE*, vol. 78, pp. 1415–1442, Sept. 1990

[120] A. N. Willson, Jr., "Limit cycles due to adder overflow in digital filters," *IEEE Transactions on Circuit Theory*, vol. CT-19, pp. 342–346, July 1972

[121] G. Yen, A. N. Michel, "A learning and forgetting algorithm in associative memories: Results involving pseudo-inverses," *IEEE Transactions on Circuits and Systems*, vol. 38, pp. 1193–1205, Oct. 1991

[122] G. Yen, A. N. Michel, "A learning and forgetting algorithm in associative memories: The eigenstructure method," *IEEE Transactions on Circuits and Systems-II: Analog and Digital Signal Processing*, vol. 39, pp. 212–225, Apr. 1992

[123] F. Zou, J. A. Nossek, "Stability of cellular neural networks with opposite-sign templates," *IEEE Transactions on Circuits and Systems*, vol. 38, pp. 675–677, June 1991

[124] *Proceedings the 1990 IEEE International Workshop on Cellular Neural Networks and Their Applications*, Budapest, Hungary, Dec. 1990

ABOUT THE AUTHORS

Derong Liu was born in Jilin, China, in January 1963. He received the B.S. degree in mechanical engineering from the East China Institute of Technology in 1982, the M.S. degree in electrical engineering from the Institute of Automation, Chinese Academy of Sciences in 1987, and the Ph.D. degree in electrical engineering from the University of Notre Dame in 1993. During his first year of graduate study at Notre Dame (1990–1991), he received the Michael J. Birck Fellowship.

From 1982 to 1984, he worked as an electro-mechanical engineer at China North Industries Corp., Jilin, China. From 1987 to 1990, he was a faculty member in the Department of Electrical Engineering at the Graduate School of Chinese Academy of Sciences, Beijing, China. Since October 1993, he is with the Electrical and Electronics Research Department, General Motors NAO R&D Center. His research interests include systems and control theory, fault diagnosis, signal processing, and neural networks. He is a member of Eta Kappa Nu.

Anthony N. Michel was born in Rekasch, Romania, in November 1935. He received the B.S. degree in electrical engineering, the M.S. degree in mathematics, and the Ph.D. degree in electrical engineering from Marquette University, Milwaukee, WI, and the D.Sc. degree in applied mathematics from the Technical University of Graz, Austria.

Dr. Michel has seven years of industrial experience and has held positions with Stearns Magnetic Products, Milwaukee, WI, the U.S. Army Corps of Engineers, and A. C. Electronics, a Division of GM, Oak Creek, WI. In 1968, he joined the faculty of Iowa State University, Ames, where he was a Professor in the Department of Electrical Engineering until 1984. From 1984 to 1988, he was Frank M. Freimann Professor and Chairman of the Department of Electrical Engineering, University of Notre Dame, Notre Dame, IN. Currently, he is Frank M. Freimann Professor and Matthew H. McCloskey Dean of the College of Engineering at the University of Notre Dame. He is coauthor of the books *Qualitative Analysis of Large Scale Dynamical Systems* (with R. K. Miller) (New York: Academic Press, 1977), *Mathematical Foundations in Engineering and Science: Algebra and Analysis* (with C. J. Herget) (Englewood Cliffs, NJ: Prentice-Hall, 1981), and *Ordinary Differential Equations* (with R. K. Miller) (New York: Academic Press, 1982).

Dr. Michel received the 1978 Best Transactions Paper Award from the IEEE Control Systems Society, the 1984 Guillemin-Cauer Award from the IEEE Circuits and Systems Society, and an IEEE Centennial Medal. He is a former Associate Editor and a former Editor of the IEEE TRANSACTIONS ON CIRCUITS AND SYSTEMS, a former Associate Editor of the IEEE TRANSACTIONS ON AUTOMATIC CONTROL, and a former Associate Editor of the IEEE TRANSACTIONS ON NEURAL NETWORKS. He was Co-Chairman of the Organizing Committee of the 1978 Midwest Symposium on Circuits and Systems, he was the Program Chairman of the 1982 IEEE International Large Scale Systems Symposium, he was the Program Chairman of the 1985 IEEE Conference on Decision and Control, he was President of the IEEE Circuits and Systems Society in 1990, and he was Co-Chairman of the 1990 IEEE International Symposium on Circuits and Systems. He is an Associate Editor at Large for the IEEE TRANSACTIONS ON AUTOMATIC CONTROL, and he is an elected member of the Board of Governors in the IEEE Control Systems Society.

Dr. Michel was a Fulbright Scholar in 1992 (at the Technical University of Vienna, Austria). He is a Foreign Member of the Academy of Engineering of the Russian Federation.

Dr. Michel is a member of Pi Mu Epsilon, Eta Kappa Nu, Phi Kappa Phi, and Sigma Xi, and a Registered Professional Engineer (in the State of Wisconsin).

Lecture Notes in Control and Information Sciences

Edited by M. Thoma

1989–1993 Published Titles:

Vol. 178: Zolésio, J.P. (Ed.)
Boundary Control and Boundary Variation.
Proceedings of IFIP WG 7.2 Conference,
Sophia- Antipolis,France, October 15-17,
1990.
392 pp. 1992 [3-540-55351-7]

Vol. 179: Jiang, Z.H.; Schaufelberger, W.
Block Pulse Functions and Their Applications in
Control Systems.
237 pp. 1992 [3-540-55369-X]

Vol. 180: Kall, P. (Ed.)
System Modelling and Optimization.
Proceedings of the 15th IFIP Conference,
Zurich, Switzerland, September 2-6, 1991.
969 pp. 1992 [3-540-55577-3]

Vol. 181: Drane, C.R.
Positioning Systems - A Unified Approach.
168 pp. 1992 [3-540-55850-0]

Vol. 182: Hagenauer, J. (Ed.)
Advanced Methods for Satellite and Deep
Space Communications. Proceedings of an
International Seminar Organized by Deutsche
Forschungsanstalt für Luft-und Raumfahrt
(DLR), Bonn, Germany, September 1992.
196 pp. 1992 [3-540-55851-9]

Vol. 183: Hosoe, S. (Ed.)
Robust Control. Proceesings of a Workshop
held in Tokyo, Japan, June 23-24, 1991.
225 pp. 1992 [3-540-55961-2]

Vol. 184: Duncan, T.E.; Pasik-Duncan, B.
(Eds.)
Stochastic Theory and Adaptive Control.
Proceedings of a Workshop held in Lawrence,
Kansas, September 26-28, 1991.
500 pages. 1992 [3-540-55962-0]

Vol. 185: Curtain, R.F. (Ed.); Bensoussan, A.;
Lions, J.L.(Honorary Eds.)
Analysis and Optimization of Systems: State
and Frequency Domain Approaches for Infinite-
Dimensional Systems. Proceedings of the 10th
International Conference, Sophia-Antipolis,
France, June 9-12, 1992.
648 pp. 1993 [3-540-56155-2]

Vol. 186: Sreenath, N.
Systems Representation of Global Climate
Change Models. Foundation for a Systems
Science Approach.
288 pp. 1993 [3-540-19824-5]

Vol. 187: Morecki, A.; Bianchi, G.;
Jaworeck, K. (Eds.)
RoManSy 9: Proceedings of the Ninth
CISM-IFToMM Symposium on Theory and
Practice of Robots and Manipulators.
476 pp. 1993 [3-540-19834-2]

Vol. 188: Naidu, D. Subbaram
Aeroassisted Orbital Transfer: Guidance and
Control Strategies.
192 pp. 1993 [3-540-19819-9]

Vol. 189: Ilchmann, A.
Non-Identifier-Based High-Gain Adaptive
Control.
220 pp. 1993 [3-540-19845-8]

Vol. 190: Chatila, R.; Hirzinger, G. (Eds.)
Experimental Robotics II: The 2nd International
Symposium, Toulouse, France, June 25-27
1991.
580 pp. 1993 [3-540-19851-2]

Vol. 191: Blondel, V.
Simultaneous Stabilization of Linear Systems.
212 pp. 1993 [3-540-19862-8]

Vol. 192: Smith, R.S.; Dahleh, M. (Eds.)
The Modeling of Uncertainty in Control
Systems.
412 pp. 1993 [3-540-19870-9]

Vol. 193: Zinober, A.S.I. (Ed.)
Variable Structure and Lyapunov Control
428 pp. 1993 [3-540-19869-5]

Vol. 194: Cao, Xi-Ren
Realization Probabilities: The Dynamics of
Queuing Systems
336 pp. 1993 [3-540-19872-5]